D0418841

Renewable Energy

Renewable Energy

A USER'S GUIDE

ANDY MCCREA

Foreword by
Dr Jeremy Leggett

THE CROWOOD PRESS

First published in 2008 by
The Crowood Press Ltd
Ramsbury, Marlborough
Wiltshire SN8 2HR

www.crowood.com

British Library Cataloguing-in-Publication Data
A catalogue record for this book is available from the British Library.

ISBN 978 1 84797 061 9

Disclaimer
The author and publisher do not accept responsibility, in any manner whatsoever, for any error, or omission, nor any loss, damage, injury, adverse outcome or liability of any kind incurred as a result of the use of any of the information contained in this book, or reliance upon it. Readers are advised to seek specific professional advice relating to their particular house, project and circumstances before embarking on any building or installation work.

Typeset by Magenta Publishing Ltd (www.magentapublishing.com)

Printed and bound in Singapore by Craft Print International Ltd

Contents

Foreword

Renewable energy markets have some big drivers behind them these days. Firstly, more and more people are suspecting that there isn't as much oil and gas left as we are being told. This problem has become known as 'peak oil'. Crudely speaking, this is the time when all the oil discovered, and to be discovered, is half gone: the time when oil supply stops growing, starts becoming increasingly unaffordable, and sets us on the road to a world where only the super-rich can afford to drive and fly.

Faced with this possibility, some want to turn to the Canadian tar sands and to coal. In Canada, there is a lot of tar sand that, after much burning of gas to melt the tar, can be turned into oil. In other countries, there is plenty of coal and, after much emission of gas, that too can be turned into oil.

Oil companies are trying to turn themselves into tar and coal companies. This is a big mistake. To do this, they need to forget about a second mega-problem that is going to change our lives: global warming.

The thermostat of our planet is in danger of running out of control. The impacts of this global overheating are snowballing. Economies are being ruined (think of the Mexican floods in Tabasco province). Living systems are being decimated (think of coral bleaching). Lives are being turned on end (think of Hurricane Katrina, the Australian drought, and so on). Much worse is to come unless we change.

In 1939, when Winston Churchill said 'we are entering a time of consequences', he wasn't kidding. The British face the same magnitude of threats now as we did then. Invasion, of a kind. Ruin, just as bad as bombing. Not just the Brits. Everyone.

We all have to mobilize as though for world war, just as in 1939. Then – in a big hurry – we made bombers, fighters and tanks faster than you would have believed possible. But in the war we've got to fight today, we need different weapons.

First, we need energy conservation of all kinds. Second, we need energy efficiency of all kinds. Third, we need renewable energy of all kinds.

This imperative is what makes books like Andy's so important. We need people to have faith that renewables can deliver, if we just but give them a chance. By marrying renewable energy with energy conservation and efficiency in our homes, office and factories we stand a chance. Without making these changes we really don't ... and time is running out.

Jeremy Leggett, Executive Chairman, Solarcentury

DEDICATION

To my children, John and Suzanne, and the future generations who will live in the energy-hungry world of the future.

To the staff of *Action Renewables*, my friends and colleagues who have given me so much encouragement and support.

To my wife, Shirley, without whom this would not have been possible.

ACKNOWLEDGEMENTS

Thanks are due to my many colleagues and the organizations, too many to mention individually, who offered and provided information and images used throughout this book. The captions associated with the images and figures give the name/type of the product/system/service and the name of the company that manufactures/supplies or operates it, where appropriate.

The depiction of, and reference to, specific products/systems/services should not be taken as endorsement of them and no responsibility can be accepted for the subsequent use of such.

Acknowledgements are also made, where relevant, to other sources from which images and information may have been adapted or taken. Details of the relevant organizations, contacts and companies are given in the Useful Information section at the end of each chapter.

CHAPTER 1

Renewable Energy: An Introduction

To ensure that we move rapidly to a low carbon economy embracing the concept of zero carbon housing, it will be necessary to build super-insulated dwellings and integrate a range of renewable technologies at the design and construction stage. We have no hope of achieving ambitious targets for carbon reduction (60 per cent by 2050) without recourse to bold initiatives.

RENEWABLE ENERGY – NEW ENERGY?

Renewable energy, despite the inference in the name, is anything but new. Indeed, most of the technologies now involved have been in the service of man for a very long time. The current thirst to understand more about these technologies has been fuelled by a range of imperatives, perhaps most significantly a keener awareness that we have been abusing the Earth's resources for too long and the growing anticipation that dire consequences are now just about to arrive. The perception is that the world's weather has become more unpredictable and extreme, with almost daily reports of flooding, hurricanes, tornados, tsunamis and record-breaking temperatures – some of it on our very doorstep and much of it being blamed directly (rightly or wrongly) on climate change.

The sun, the wind and water power have been lightening the load of mankind for many centuries, but it is only recently that their potential has been pursued with unprecedented vigour. It always seems that our continental neighbours, in countries like Denmark, Germany and Sweden, are ahead in progress towards the necessary sustainable solutions, but recent advances and new government thinking on energy have made these renewable technologies much more accessible closer to home.

At first sight renewables appear to offer a utopian solution to the problems of modern energy demand, but everything comes at a price and their wide-scale deployment is anything but straightforward. For most governments of the world, fuel security and protection of the environment is now in the front line of policy and recently 'green' credentials have become vote capital on the hustings, with leaders from the major political parties trying to gain the edge by posturing as more environmentally friendly than their opponents.

There is also something inbuilt in each of us that recognizes that the environment is precious and its despoilment is wrong – even though we continue to demand more and more energy with apparent disregard for the consequences. As we shall see, one challenge of the zero carbon house concept is that the demand for electricity to drive our ever more sophisticated electronic appliances and entertainment systems is rising steadily. It is a fact of life that energy is central to our modern civilization but concerns about climate change, combined with rising energy costs and a fear of uncertainty around future energy supplies, have fuelled the drive for greater delivery of renewable energy. We all now aspire to minimize our carbon footprint and hope for a sustainable energy future that will provide us with the standard of living we have come to accept, but which will not compromise that of future generations. Renewable technologies appear to offer a partial solution, providing both low carbon heat and electricity (on the shoulders of conservation and energy efficiency) to help us move forward on this objective.

The aim of this book is to bring greater insight to some of the huge number of people who want to know more about renewable energy and how they can get involved with it. Over the last couple of decades the idea of making electricity and heat for our homes using renewable energy has moved from being a fringe, often DIY, pursuit to a mainstream business worth billions to the world's economy. Much of this advance has been as a result of government support through a variety of grant programmes and through the awareness-raising activities of organizations involved in this area. I have called the book 'a user's guide' rather than 'a beginners' guide', as my original intention had been, because there is a higher level of awareness now about the issues (and the devices) and this allows me to move directly to the technologies themselves and review them in a little more detail. There is also a good deal of interest, fuelled by the media (and the weather!), in the whole area of climate change. It seems that people genuinely want to know how they can do their bit to improve their carbon footprint from a size 12 to a size 7. The technologies involved are now at a reasonable level of sophistication and the proper design and installation of renewable energy systems in the home or in businesses can deliver effective, reliable, affordable heat and electricity.

WHAT IS RENEWABLE ENERGY?

The term 'renewable energy' is frequently confused with 'recycling' but, despite the tenuous links, the two are different. Renewable energy may be defined in various ways that sometimes include different technologies (hydro, for example, is often omitted). A suitable definition might be that renewable energy is obtained from sources that are essentially inexhaustible (or that may be replenished), unlike, for example, the fossil fuels, of which there is a finite supply. In particular, they emit no greenhouse gases or are 'emissions neutral' over their lifetimes. They are also often referred to as 'energy sources replenished by natural phenomena' (this excludes any use of nuclear fuels). For the purposes of this book, renewable sources of energy are wood (bioenergy), some wastes, deep-geothermal energy, wind, solar photovoltaic (PV) and solar thermal energy (collected using solar panels and heat pumps). More recently, marine energy from the oceans, such as wave and tidal energy, have been included. Heat pumps use heat energy from a range of sources that, relatively speaking, are not particularly warm – for example the air, the ground (a couple of metres below the surface), lakes, rivers or even the sea. This is referred to as 'low-grade heat' and it requires to be 'pumped-up' using electricity to heat water (or air) to a more useful temperature (30–40°C). The same heat pumps may also be used (in a different configuration) to provide space cooling if necessary.

Associated terms are in common usage, such as microgeneration, embedded generation and alternative energy. In this book I will stick to 'renewable energy' and will address the principal heat and electricity technologies noted above. Microgeneration is

Figure 1: Whisper Tech WhisperGen Domestic CHP. At the heart of the domestic CHP unit is a Stirling engine, controlled by a battery of microcomputers, which produces space heating and electricity. The model shown is fuelled by natural gas and produces around 8kW of heat and 1.1kW of electricity. Work is in progress to develop similar CHP engines that can operate on a range of fuels including biomass. (Author/ www.whispergen.com)

The Greenhouse Effect

The 'greenhouse effect' produces a rise in temperature in the atmosphere because certain gases present, known as greenhouse gases (water vapour, carbon dioxide, nitrous oxide and methane), trap the energy from the Sun. Incoming solar radiation passes almost unimpeded through the Earth's atmosphere and eventually it reaches the oceans or the Earth's surface. This radiation is absorbed and then re-radiated at longer wavelengths as heat (or infrared radiation). The greenhouse gases in the atmosphere, however, do not allow this outgoing infrared radiation to escape; instead the heat is trapped and it raises the temperature of the atmosphere. The greenhouse effect is an entirely natural process. Without these gases the heat would be lost into space, the Earth's average temperature would be about 30°C colder and life as we know it would not exist.

The consensus of the world's climate scientists, represented by the IPCC, is that the quantity of greenhouse gases present in the atmosphere, notably carbon dioxide, has been increasing steadily over the last 250 years and it now far exceeds the natural range of the past 650,000 years. The current level of carbon dioxide is estimated to be around 370 parts per million (ppm). It is almost certain that this rapid increase is attributable to man, most of it as a result of increasing industrial and economic growth, and that the rising atmospheric concentration of carbon

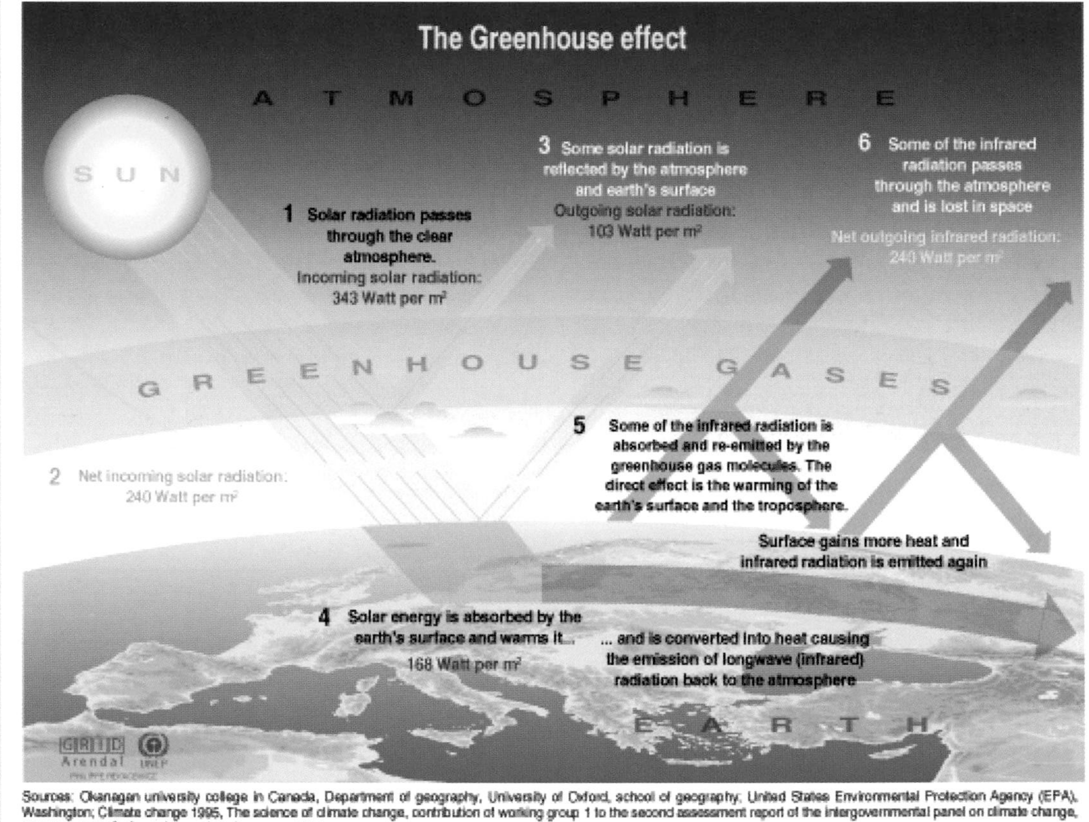

Figure 2: Human activities are causing greenhouse gas levels in the atmosphere to increase. This graphic explains how solar energy is absorbed by the Earth's surface, causing the Earth to warm and emit infrared radiation. The greenhouse gases then trap the infrared radiation, thus warming the atmosphere. (Philippe Rekacewicz)

dioxide and the other gases is increasing the temperature in the atmosphere.

There is an ongoing debate about how much of an increase can be sustained before we reach the 'tipping point', the point of no return after which positive-feedback processes are instigated, releasing even greater quantities of various greenhouse gases into the atmosphere. This rise in temperature will eventually unlock methane from the permafrost layers and release vast amounts of carbon dioxide from the oceans. As the polar ice sheets melt, there will be a resulting increase in sea level. There is uncertainty about the exact timing and consequences of reaching the tipping point, but there can be no doubt that, if this is a realistic scenario, the effect on the global population will be widespread and catastrophic.

It has been suggested by the IPCC that a global rise in average temperature of 2°C above the level before the Industrial Revolution in the late eighteenth century, or an increase in the carbon dioxide level in the atmosphere to around 450ppm (from the current 370ppm) would be sufficient to trigger the tipping point. In its last major report in 2001, the IPCC predicted a rise in global temperatures of 1.4°C to 5.8°C between 1990 and 2100, but the estimate only takes account of global warming driven by known greenhouse gas emissions. Evidence suggests that we are currently contributing around 2ppm of carbon dioxide a year and rising, and can therefore expect that the tipping point will reach us within thirty years. Internationally agreed targets have been set to reduce the quantities of gases being emitted. The targets cover emissions of the six main greenhouse gases:

- Carbon dioxide (CO_2)
- Methane (CH_4)
- Nitrous oxide (N_2O)
- Hydrofluorocarbons (HFCs)
- Perfluorocarbons (PFCs)
- Sulphur hexafluoride (SF_6).

In November 2007 the IPCC reported in its *Draft Synthesis Report* that 'Warming of the climate system is unequivocal, as is now evident from observations of increases in global average air and ocean temperatures, widespread melting of snow and ice, and rising global average sea level.'

a specific term defined in the UK Energy Act, 2004, as the generation of electricity up to 50 kilowatts (kWe), or by the production of heat up to 45 kilowatts thermal (kWth), by any plant that relies wholly or mainly on a source of energy or a technology in the following list: biomass, biofuels, fuel cells, photovoltaics, water (particularly small-scale hydroelectric plant), wind, solar power, ground/air heat pumps, and combined heat and power systems (CHP). A domestic CHP or microgeneration unit, as shown on page 9, is designed to provide heat and electricity using a natural gas-fuelled Stirling engine. The hope is that these will become economically viable and in widespread use. Eventually they may use biomass fuels, such as wood pellets. In January 2008 British Gas (Centrica) contracted with CERES Power to develop wall-mounted, fuel cell-based CHP microgeneration units through field proving trials.

CLIMATE CHANGE

Now that we have a clearer view of what is meant by renewable energy, it is necessary to understand why it should be used in place of the conventional sources. Climate change, specifically the rise in global temperature, has emerged as one of the most important issues facing our modern world and we are constantly informed that urgent, corrective control over the amount of greenhouse gases we produce needs to be exercised. If action is not taken soon, according to the Intergovernmental Panel on Climate Change (IPCC), many millions of people will be at risk from extreme events such as heat waves, drought, floods and storms. Coasts and cities will be threatened by rising sea levels and ecosystems, plants and animal species will be in imminent danger of extinction.

The main greenhouse gas causing concern in relation to energy is carbon dioxide (or CO_2). Significant amounts of CO_2 are released into the atmosphere in the course of heating our homes and businesses, and in generating electricity through the burning of fossil fuels (coal, oil and natural gas) in power stations. Perhaps the most flexible and amenable energy source is electricity, which is not a fuel. Despite the overwhelming view of the world's climate scientists that we are currently in a period of rapid climate change, a number of sceptics remain convinced that the observed increase in global temperature is due to a combination of other factors. Regardless of the truth, the indisputable fact is that fossil fuel resources are diminishing rapidly, while their transportation and combustion pollute and poison the air, sea and land around us. The rising demand for the world's decreasing fossil fuel resources has led to wars, political instability and territorial disputes that will most likely increase in the years ahead.

The wider introduction of small-scale renewables and microgenerators may not initially have much impact on national targets, but in time it will significantly affect how we use (and waste) energy in our homes and will help deliver the paradigm shift that provides the low carbon future. The generation of electricity and heat in our homes, businesses and communities using small-scale renewable energy will help to reduce emissions of carbon dioxide, and at the same time minimize energy losses associated with the transmission of electricity from power stations to its point of use (transmission losses are between 5 and 10 per cent). In future it is hoped that super-insulated buildings will have dramatically lower heat requirements than current buildings and that these can be met from renewable heat technologies. Eventually electricity-producing technologies, such as photovoltaic panels and small-scale wind turbines integrated within a building's structure, will fully supply the demand from lighting, cooking and entertainment appliances.

In recent years another factor, security of fuel supply, has emerged as important in the overall consideration of our energy future. Despite recent finds, the on- and offshore resources of oil and gas around the British Isles are past their peak production and we are no longer in a position to support our energy requirements from these reserves. This means that imported fuels are necessary and we must compete on international markets for these vital commodities. The future global availability of these limited resources and the security of their supply will depend on factors outside our direct control and will rely on the stability of producer countries. The likely sustained increases in the cost of fossil fuels, combined with the growing uncertainties surrounding their supply, are additional reasons for pursuing a renewable energy strategy as part of any energy policy.

PUTTING RENEWABLE ENERGY INTO ENERGY POLICY

The 1997 Kyoto Protocol to the United Nations Framework Convention on Climate Change is recognized as the most important legislation driving climate change policy over the last decade. The Protocol itself is an amendment to the International Treaty on Climate Change, assigning mandatory emission limitations for the reduction of greenhouse gas emissions to the signatory nations. The objective of the Protocol is the 'stabilization of greenhouse gas concentrations in the atmosphere at a level that would prevent dangerous anthropogenic interference with the climate system'. The Kyoto Protocol commits countries and member states to individual, legally binding targets to limit or reduce their greenhouse gas emissions; it came into effect in February 2005 and thirty-four industrialized nations are obligated to reduce their greenhouse gas emissions by 5.2 per cent below 1990 levels by 2008–12. As of June 2007, a total of 172 countries and other governmental entities have ratified the agreement (representing more than 61.6 per cent of global emissions). Notable exceptions included the United States and Australia. Other countries that have ratified the Protocol, including India and China, were not required to reduce carbon emissions under the Protocol.

In December 2007 the Indonesian government brought representatives from more than 180 countries together with observers from intergovernmental and non-governmental organizations to the island of Bali to focus on the next round of negotiations for a new global climate treaty. In what has become known as the 'Bali Road Map', the prime objective was to get a unilateral commitment to limit global warming to no more than 2°C above the pre-industrial temperature; this limit has already been formally adopted by the European Union and a number of other countries. The Bali Climate Change Conference set down terms of reference for the future Copenhagen Protocol ('Kyoto Phase 2') that were based on the best available science. These called for the long-run greenhouse gas concentrations to be stabilized at a level well below 450ppm, through a reduction in

United Kingdom Government Climate Change Bill

In March 2007 the UK government committed in law to reducing carbon dioxide emissions. The Bill has the following provisions:

- Britain is to become the first country to set legally binding carbon dioxide emission reduction targets. CO_2 will be cut by between 26 per cent and 32 per cent by 2020 and 60 per cent by 2050.
- There will be a new system of five-year carbon budgets to cap total emissions. Limits will be set fifteen years in advance to assist with business planning. The setting of caps will allow a trajectory to be plotted with the aim of hitting longer-term CO_2 targets.
- Courts *will* be given powers to name and shame government ministers if targets are missed.
- An independent committee on climate change will be established to monitor and advise on progress.
- Annual reports to Parliament will be made, detailing progress towards the emission targets.
- Ministers will be required to produce five-year reports on the potential impact of climate change and their proposed actions.
- New powers will be granted to allow government ministers to impose future controls on emissions, such as a possible domestic emissions trading scheme.

global greenhouse gas emissions of at least 50 per cent below their 1990 levels by the year 2050. The Bali Climate Change Conference is to be followed by two subsequent meetings at Pozna in Poland (2008) and Copenhagen in Denmark (2009), where the details of the binding national commitments will be agreed.

EU AND UK RENEWABLE ENERGY POLICY

One of the principal drivers for the greater deployment of renewables across Europe is the European Commission's energy policy. Renewable energy – particularly that from wind, water, solar power and biofuels – is a central pillar of this policy. As well as achieving an increase in energy sustainability and securing a significant reduction in carbon dioxide, the policy will also address the increased reliance of the European Community on imported fossil fuels.

In January 2008 European Union (EU) leaders mandated their member states to increase their uptake of renewable energies in an effort to boost the EU's share from 8.5 per cent in 2008 to 20 per cent by 2020. A separate target to increase EU biofuels use to 10 per cent of transport fuel consumption is to be achieved by every country as part of the overall objective. To achieve these aims, each of the twenty-seven nations in the community is required to increase its share of renewables by 5.5 per cent from 2005 levels, with the remaining increase calculated on the basis of per capita gross domestic product (GDP).

Countries can decide their preferred 'mix' of renewables in order to take account of their different renewable energy potentials, but they must present national action plans (NAPs) to the Commission by 31 March 2010. These plans will detail national strategies for three sectors: electricity, heating and cooling, and transport. The United Kingdom is required to increase its share of renewable energy to 15 per cent and Ireland's objective is set at 16 per cent by 2020. Interim targets have been declared by the Commission to ensure steady progress towards the legally binding 2020 target.

In 2000 the British government announced the UK Climate Change Programme (CCP), which confirmed a domestic policy goal of moving towards a reduction in emissions of carbon dioxide by 20 per cent below 1990 levels by 2010. In 2003 an Energy White Paper further developed the policy by declaring four main objectives:

- Tackling climate change and greenhouse gas emissions and putting the UK on a path to cut CO_2 emissions by 60 per cent by 2050.
- Maintaining the reliability of energy supplies.
- Promoting competitive markets.
- Ensuring every home is adequately and affordably heated.

The programme was updated in spring 2006. This was followed in July 2006 by the publication of another review of energy policy, *The Energy Challenge*, intended to examine progress against the medium- and long-term 2003 Energy White Paper goals and consider options for further steps to achieve them. The whole debate was given greater urgency in October 2006 when Sir Nicholas Stern published his *Review of the Economics of Climate Change*, the most comprehensive review on this subject to date. The first half of the review focused on the impacts and risks arising from uncontrolled climate change and the associated costs and available opportunities to tackle it; the second half examined the national and international policy challenges of moving to a low carbon economy. The British government recognized the valuable contribution that small-scale generation can offer and consulted on a national microgeneration strategy. The objective of this consultation was to create conditions under which microgeneration, including renewable technologies, could offer a realistic alternative or become a supplementary generation source for the householder, community and business.

A very significant step forward was taken in March 2007 when the British government became the first in the world to commit itself to legally binding reductions in carbon dioxide emissions in a Climate Change Bill. Along with an accompanying strategy, the Bill set down a framework for moving the UK to a low carbon economy. It also demonstrated the UK's commitment as progress continues towards establishing a post-2012 global emissions agreement. The bill will enshrine a legal commitment (as set down previously in the 2003 White Paper) to cut emissions by 60 per cent by 2050.

RENEWABLES OBLIGATION

The Renewables Obligation (RO) for England and Wales and the associated Renewables (Scotland) Obligation, which came into force in April 2002, were introduced as a direct incentive to encourage the uptake of new renewables generation. The Northern Ireland Renewables Obligation (NIRO) was introduced in April 2005. These place a mandatory requirement for UK electricity suppliers to source a growing percentage of electricity from eligible renewable generation capacity (currently increasing to 15 per cent by 2015). Eligible renewable generators receive Renewables Obligation Certificates (ROCs) for each MWh (1,000 units) of electricity generated. These ROCs can then be sold to suppliers, in order to fulfil their obligation.

Suppliers are required to produce evidence of their compliance with the Obligation to the Office of Gas and Electricity Markets (Ofgem) by presenting ROCs. Suppliers can either present enough certificates to cover the required percentage of their output, or they can pay a buyout price for any shortfall. All proceeds from buyout payments are recycled to suppliers in proportion to the number of ROCs they present. The buyout price is set each year by Ofgem (in 2007/8 it stands at £34.30/MWh). ROC trading is administered by the Non-Fossil Purchasing Agency Ltd. ROCs have increased the profitability of renewable energy generation as the certificates have an additional value over and above the price of electricity itself. This is especially true for wind, which is already generating electricity at prices competitive with thermal-fired power plant. The value of a technology's output can be emphasized by increasing the number of ROCs it receives per MWh generated. In the UK it has been proposed that marine technologies can be encouraged by awarding them double ROCs.

RENEWABLE ENERGY DOMESTIC AND COMMUNITY GRANTS

In the UK the uptake of renewable energy systems for domestic, community and other buildings was greatly encouraged by the introduction of the Clearskies government grants in 2003. In 2006 the government replaced Clearskies with the Low Carbon Building Programme (LCBP). This programme underlined the importance of energy efficiency by requiring homes to be insulated and have energy-efficient lighting and controls fitted before a renewable energy grant could be awarded. Six renewable energy technologies are supported at around 30 per cent of the installed cost. The Scottish Community and Householder Renewables Initiative (SCHRI) provides grants for communities and householders in Scotland. Under the Northern Ireland Environment and Renewable Energy Fund (EREF), an £8 million household grants programme was launched in 2006. Known as Reconnect, the programme also supported six renewable technologies with grants between 30 and 50 per cent of the installed cost. Due to the difficulty in estimating uptake in these schemes, the duration and available funding is subject to change at any time. Contact details for these schemes are given at the end of the chapter (page 25).

POLICY IN IRELAND

As noted, the European Union White Paper target calls for the member states, including Ireland, to achieve a 12 per cent contribution from renewable energy sources (sometimes known as RES) to the EU's Total Primary Energy Requirement (TPER) by 2010. Over the last couple of decades, Ireland has experienced a period of strong economic growth and this has meant that its contribution from renewables has fallen below the average of the EU-15 countries (the fifteen countries that made up the Union before its enlargement). One of the key strengths of renewable energy is its indigenous nature. The fall in output from the gas fields around Ireland (Kinsale and Ballycotton) has meant that almost all the fossil fuel now consumed has to be imported, generally from outside the EU. Ireland's dependency on fossil fuels has increased from 65 per cent in 1990 to 89 per cent in 2003. The average energy import dependency of the EU-25 (the 25 countries that comprise the enlarged EU) in 2004 was around 50 per cent.

On 1 May 2006 the Minister for Communications, Marine and Natural Resources announced the official launch of Ireland's Renewable Energy Feed In Tariff (REFIT) scheme. REFIT provides support to large-scale renewable energy projects over a fifteen-year period. The new support mechanism is a change from the previous Alternative Energy Requirement (AER) programme in that it is a fixed feed in tariff rather than a competitive price tendering process. Under REFIT, applicants must have planning permission and a grid connection offer for their projects and they will then be able to contract with any licensed electricity supplier up to the notified fixed price tariffs (Table 1).

Table 1: Fixed price tariffs (prices as of January 2008).	
Large wind energy (over 5 Megawatts)	> 5.7 Euro cent per kilowatt hour
Small wind energy (under 5 Megawatts)	> 5.9 Euro cent per kilowatt hour
Biomass (landfill gas)	> 7.0 Euro cent per kilowatt hour
Hydro and other biomass technologies	> 7.2 Euro cent per kilowatt hour

For small-scale renewables the Irish government announced the Greener Homes scheme, which was allocated €27 million in the 2006 budget package. The 2007 budget provided for an increase of €5 million and provision for additional spending of €7 million in 2008 and €8 million in 2009. The scheme provides grant assistance to homeowners who intend to install a renewable energy heating system for either new or existing homes. The scheme is administered by Sustainable Energy Ireland (SEI) and provides support for three heating technologies, solar water heating, heat pumps and biomass.

In March 2007 the Irish government published an energy White Paper, *Delivering a Sustainable Energy Future for Ireland*, which describes proposed actions and targets for the energy policy framework out to 2020. The Paper sets a clear path for meeting the government's goals of ensuring safe and secure energy supplies, promoting a sustainable energy future and supporting competitiveness, and it enshrines the objectives of a lower carbon energy mix with renewable energy playing a major role.

Table 2: Heat loss from a typical two-storey dwelling.	
Heat Loss	**Heat Lost (%)**
Walls	37
Roof	13
Ventilation and draughts	20
Doors and windows	19
Floors	11

ENERGY EFFICIENCY AND HEAT LOSS IN BUILDINGS

Most of the existing dwellings in the British Isles, especially those constructed before the 1970s, suffer from huge heat loss. Installing high efficiency or renewable energy based heating systems will be of little value if the precious heat leaves the building almost as soon as it arrives. The design standards and building fabric of our homes differ dramatically from those of our continental neighbours, especially Scandinavia, where building insulation standards are much higher. Before any form of renewable heating is even contemplated, it is essential that the building is insulated to the highest level – above the building control standards, where possible. There are two main causes of heat loss from our buildings: heat conduction through the walls, windows, floor and roof, and uncontrolled ventilation heat losses through open or draughty windows and doors. Table 2 shows how, according to figures from the Energy Saving Trust, heat is lost from a typical two-storey dwelling.

From this it can be seen that walls represent the largest source of heat loss in a building. Cavity wall insulation (CWI) is an excellent way to reduce this. In buildings with no cavities, or cavities that cannot be filled, it may be possible to apply external insulation. Walls that are insulated on their internal surfaces will lose some of their thermal inertia, their ability to retain the Sun's heat, which has accumulated throughout the day and would otherwise be re-radiated into the house. New buildings can benefit from prefabricated, highly insulated wall panels and from stricter Building Control Regulations demanding

Figure 3: The Kingspan 'Lighthouse' net-zero carbon home. The home achieves CSH Level 6, the standard to which all new homes should be designed and constructed by 2016. (Image reproduced courtesy Kingspan Century)

higher standards for insulation and construction fabric. Best practice would suggest that constructors should go for insulation standards around 25 per cent above those recommended in the Building Control Regulations. Roof space insulation is usually straightforward to apply and can be mineral wool, paper or polystyrene laid to a thickness of at least 270 to 300mm.

U-values

The U-value, or thermal transmittance, is a measure of the rate at which heat passes through (or is lost through) a building fabric or component (window, door, roof, floor, etc.) in W/m^2K. Each material used in a building's construction will have a different U-value, depending on its insulating properties (lower U-values are best). U-values can be calculated knowing the thicknesses and thermal conductances of the individual materials that make up any building component. Previously, the overall U-value for a property was calculated by totalling the individual component U-values and this summation had to be lower than a target value to comply with Building Control Regulations. This method, known as the calculated trade-off, has now been superseded by a new method of demonstrating compliance known as the whole-building approach or the whole-building method.

Homes built in the British Isles before and during the 1970s had virtually no insulation, although the situation improved as the Building Control Regulations were reviewed. New homes are now to be super-insulated, that is they will have levels of insulation for the building fabric, roof, floors, windows and doors well above that specified even in the most recent version of the Building Control Regulations. Insulation is cheap and simple to install at the construction stage and it provides fuel savings year on year. It is more difficult to retrofit insulation into existing properties, especially the floors.

Building Control Regulations typically call for a U-value of around $0.3W/m^2K$. A super-insulated new dwelling could have an overall U-value of around $0.2W/m^2K$ and this would mean that only a small amount of heat will be required to keep the property at the desired comfort level. This infers that a smaller boiler could be installed with a consequent offset against the cost of the extra insulation, as well as the year on year savings in fuel bills. Future revisions of the Building Control Regulations will require the inclusion of renewable energy to provide the CO_2 reductions and this is discussed below in the context of the zero carbon home.

Table 3: Characteristics of some available insulation materials.

Insulating Material	Characteristics
Expanded polystyrene (EPS)	A rigid board available in various thicknesses for pitch roofs, cavity fill and floors. 'Grey' type has slightly better properties due to a graphite barrier.
Extruded polystyrene	Styrofoam has better thermal performance and moisture resistance than EPS. Contains halogens, which are environmentally unacceptable.
Mineral wool/ Mineral fibre	Glass fibre, mineral or rock wool provides flexible blankets of insulation. Drawback is the irritant fibres. Fitting and handling require special clothing and protective masks and goggles.
Polyeurothane/ Polyisocyanurate	Good insulation properties, sometimes backed with foil to enhance moisture and insulation quality. Drawback is expense, although it is useful in situations where performance is important.
Cellulose fibre	Recycled newspaper gives a sound environmental alternative that can be applied dry or sprayed onto materials wet. Excellent resistance to fire, insects and vermin.
Natural wool	Sheeps' wool, although a fibre-based material, is non-irritant and does not require special clothing or protective masks and goggles.
Cork	An excellent renewable insulation source and one that is becoming increasingly affordable as its conventional uses fall away.

Table 4: Rule of thumb heat load for a range of buildings.

Building Type	Heat Load
Existing buildings	Over 75W/m^2
Existing buildings with cavity wall insulation and roof space insulation	Around 75W/m^2
New buildings	Around 50W/m^2
Low-energy buildings (super-insulated)	30W/m^2 or below

CALCULATING A BUILDING'S HEAT LOAD

When installing any heating system, regardless of whether or not it is a renewable technology, it is important to perform a heat load or heat loss calculation. The calculation will inform just how good the building is at holding onto its heat. The heat load for new buildings is determined according to the method set down in EN 128311 ('Heating Systems in Buildings – Method for Calculation of the Design Heat Load'). There are a number of other methods that allow estimates to be made, including computer programs (see below). Clearly heat demand will vary from season to season and between individual years, so the heating requirement should be estimated on the basis of the maximum heat load. An assessment can be made for existing properties from previous annual energy bills, averaged over a number of years. This will give a better assessment than any calculation based on the house area. Rule of thumb estimates have been traditionally used by plumbers and heating engineers to give a rough estimate of the heat load (Table 4).

It is worth stressing that rule of thumb assessments are simply inaccurate estimates and must not be applied when installing heat pumps into new or existing buildings: a proper heat load calculation must be carried out. Satisfactory heat pump performance is very dependent on installing the correct size of heat pump; one that is too big or too small will result in an inefficient and expensive system.

THE CHALLENGE OF THE ZERO CARBON HOME

In December 2006 the UK government made the important announcement that by 2017 every new home constructed in the UK will be a 'zero carbon home'. The requirement to design for zero carbon housing is a challenge because it implies that absolutely no carbon is released as part of the normal activities that take place in the home. This begs the question as to where the energy for heating and to run electrical appliances, so much an essential part of modern life, must come from. The first part of the answer must be conservation and super-insulation. The second part requires source(s) of heat and electricity generation that have no carbon dioxide emissions. Small micro-generators integrated into buildings spread across the country in domestic dwellings, businesses and the public estate have the potential to meet much of the demand for heat and electricity in those buildings. This means that the losses due to centralized generation, transmission and distribution can be reduced or almost eliminated, which would help defer electricity grid re-enforcements and upgrades. There are three strands to achieving the zero-carbon home target:

- Progressive tightening of Building Control Regulations over the next decade to improve energy efficiency of construction.
- The creation of a new Code for Sustainable Homes (CSH).
- Amendment to the planning legislation to account for carbon emissions.

The primary measurement of energy efficiency in the Building Control Regulations is the Dwelling Carbon Dioxide Emissions Rate (DER), an estimate of carbon dioxide emissions per square metre of floor area. Proposed revisions to the Regulations in 2010 will require a 25 per cent reduction in DER. A further review in 2013 will seek a 44 per cent improvement with zero carbon from 2016. The challenge is to bring forward creative solutions, for example the increased use of low carbon technologies such as heat pumps. The CSH was launched in 2006 to supersede the BRE EcoHomes scheme. Initially it will only apply in England, with Scotland, Wales and Northern

Home Information Packs and Energy Performance Certificates

THIS IS AN EXAMPLE REPORT AND IS NOT BASED ON AN ACTUAL PROPERTY

Section H: Energy Performance Certificate
Save money, improve comfort and help the environment

The following report is based on an inspection carried out for:

Address:	Building type:	Home	Certificate number:	XXXX
100 Any Street,	Whole or part of building:	Whole	Date issued:	XXXX
Any Town,	Assessment method:	SAP	Name of inspector:	XXXX
Anywhere, AB1 CD2	Date of inspection:	XXXX		

This home's performance ratings

This home has been inspected and its performance rated in terms of its energy efficiency and environmental impact. This is calculated using the UK Standard Assessment Procedure (SAP) for dwellings which gives you an energy efficiency rating based on fuel cost and an environmental impact rating based on carbon dioxide (CO_2) emissions.

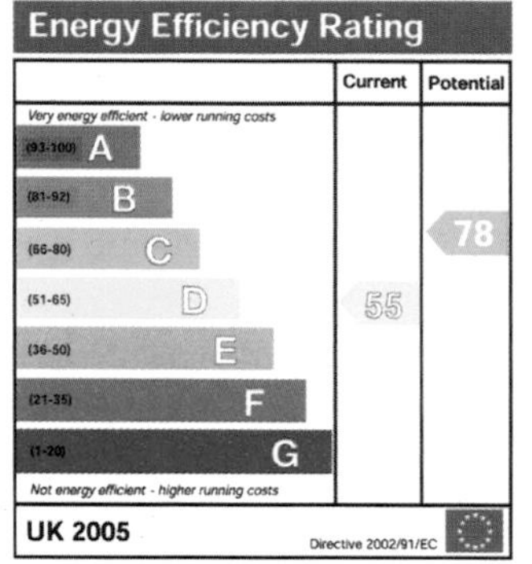

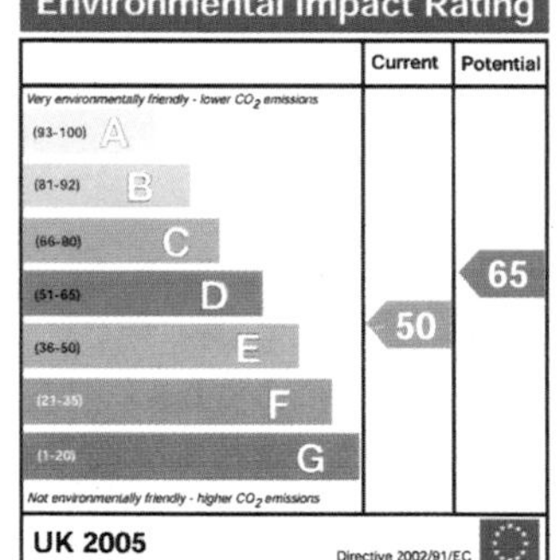

The energy efficiency rating is a measure of the overall efficiency of a home. The higher the rating the more energy efficient the home is and the lower the fuel bills will be.

The environmental impact rating is a measure of this home's impact on the environment. The higher the rating the less impact it has on the environment.

Typical fuel costs and carbon dioxide (CO_2) emissions of this home

This table provides you with an indication of how much it will cost to provide lighting, heating and hot water to this home. The fuel costs and carbon dioxide emissions are calculated based on a SAP assessment of the actual energy use that would be needed to deliver the defined level of comfort in this home, using standard occupancy assumptions, which are described on page 4. The energy use includes the energy used in producing and delivering the fuels to this home. The fuel costs only take into account the cost of fuel and not any associated service, maintenance or safety inspection costs. The costs have been provided for guidance only as it is unlikely they will match actual costs for any particular household.

	Current	Potential
Energy use	xxx kWh/m^2 per year	xxx kWh/m^2 per year
Carbon dioxide emissions	xx tonnes per year	xx tonnes per year
Lighting	£xxx per year	£xxx per year
Heating	£xxx per year	£xxx per year
Hot water	£xxx per year	£xxx per year

To see how this home's performance ratings can be improved please go to page 2

Figure 4: The front page of an EPC, a five-page report rating a home's energy efficiency and its environmental impact. Every home for sale requires an EPC as part of its Home Information Pack. Houses with solar tiles, for example, may be 'A-rated' automatically, making solar powered houses more attractive to potential buyers and tenants. (Image reproduced courtesy of Her Majesty's Stationery Office, 2008)

The EU Directive on the Energy Performance of Buildings (2002/91/EC) requires, among other things, that 'when buildings are constructed, sold or rented out, an energy performance certificate is made available to the owner or by the owner to the prospective buyer or tenant'. Energy Performance Certificates (EPCs) are an important step towards empowering homeowners to take control of their energy costs. This will be achieved by giving a clear and simple indication of the energy running costs of the building.

It is likely that the EPC will have an increasing impact on the decision-making process of the house buyer. For example, the EPC will flag up the fact that a particular house has an old, inefficient heating system, with low levels of insulation. This means that the house would be more expensive to heat than an equivalent house with good levels of insulation and a modern, efficient central heating boiler. In theory, the buyer could be in a position to ask for a reduction in the selling price based on the information highlighted in the EPC.

EPCs form part of the Home Information Pack (HIP), which is mandatory (from October 2008) in England and Wales for public sector buildings as well as homes. It is likely this requirement will be extended to Northern Ireland and Scotland. The HIP is produced by the home/building owner and includes the EPC, property deeds, searches and sales statement. The EPC will form a necessary part of constructing, selling or renting property.

In the Republic of Ireland the Building Energy Rating (BER) is an indication of the energy performance of a dwelling (similar to the EPC). It covers energy use for space heating, water heating, ventilation and lighting calculated on the basis of standard occupancy. A similar scheme to the HIPs is planned for the Irish housing market.

Table 5: CSH and Building Control Regulations energy requirements.

CSH Level	Percentage Better than DER (on 2006 Basis)	Note	Building Regulations Route Map Timescale
1	10	EST Good Practice	
2	18		
3	25	EST Best Practice	2010
4	44	Near PassivHaus	2013
5	100	Zero space/water heating and lighting	
6	Zero carbon home	Zero including cooking and appliances	2016

EST: Energy Saving Trust
PassivHaus: A specific construction standard for residential buildings that have excellent comfort conditions in both winter and summer.

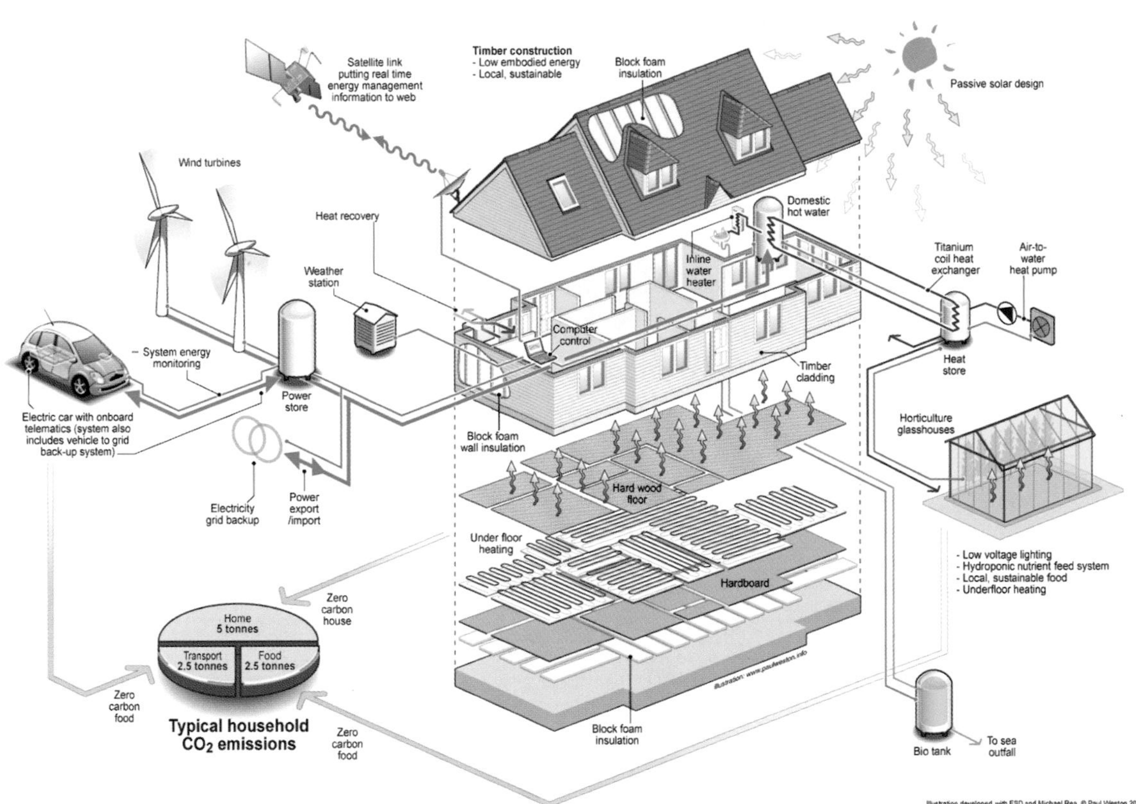

Figure 5: Zero Carbon House (ZCH) based on a low-energy demonstration project to show how renewable energy can create a unique living experience on a remote island (Shetland) in a severe climate. For more information see www.generationhomes.org.uk. (Reproduced with thanks to ESD, illustrator Paul Weston and to Michael Rea as the source of information. Copyright Paul Weston)

Power and Energy

Energy is an amount of work done or a capacity to do work, measured in Joules. ***Electricity*** is a form of energy (it is not a fuel) expressed in kilowatt hours (kWh). ***Power*** is the energy used in a given time, that is, the rate of doing work, measured in watts (W), where 1W = 1 Joule/second.

A renewable device that generates with a power of 1kW, for one hour, has produced an energy output of 1kWh (or in the case of electricity, one 'unit'). One megawatt hour (1MWh) is the same as one thousand kilowatt hours (1,000kWh). 1kWh is 3,600,000 Joules (or 3.6 Mega Joules (MJ)).

Table 6: Prefixes and multiplication factors.

Multiplication Factor	Abbreviation	Prefix	Symbol
1,000,000,000,000,000	10^{15}	peta	P
1,000,000,000,000	10^{12}	tera	T
1,000,000,000	10^{9}	Giga	G
1,000,000	10^{6}	mega	M
1,000	10^{3}	kilo	k
100	10^{2}	hecto	h
10	10^{1}	deca	da
0.1	10^{-1}	deci	d
0.01	10^{-2}	centi	c
0.001	10^{-3}	milli	m
0.000,001	10^{-6}	micro	μ

1 kilotonne (kt) = 10^3 tonnes = 1,000 tonnes
1 megatonne (Mt) = 10^6 tonnes = 1,000,000 tonnes
1 Gigagramme (Gg) = 1 kt
1 Teragramme (Tg) = 1 Mt

To convert emissions expressed in weight of carbon to emissions in weight of carbon dioxide, multiply by the ratio of their atomic masses, i.e. 44/12 (or 3.67).

Ireland still applying the EcoHomes code. The CSH allocates points in nine areas: energy, potable water, surface water run-off, materials, construction site waste/household waste, pollution, health and well being, management and ecology. Codes 5 and 6 are extremely challenging, prompting difficult questions around how zero carbon may be achieved and exactly what this means in practice (Table 5).

INTERMITTENCY AND LOAD FACTOR

Some forms of renewable energy are not available on-tap: renewable energy is only there when the wind blows, the Sun shines or the water flows. This is known as intermittency and it makes the direct comparison between different sources of energy difficult. A useful means of comparing how effectively a renewable energy technology will perform is to use its load factor. The load factor is the amount of energy (kWh) actually delivered from a generating source (fossil-fuelled or renewable) in a designated period, compared to the total possible energy that could have been delivered by the generating source system in the same time.

Using a year as the comparative period, the load factor (also referred to as capacity factor) is calculated by dividing the energy actually delivered during the year (kWh), by the peak output (in kW) over the year, and then dividing this figure by the total number of hours in a year (taken as 8,760). Instead of the peak output, the 'name plate rating' of the renewable generator may be used.

Figure 6: The Stewart Milne Group has two four-storey homes on the Building Research Establishment (BRE) Innovation Park. This building embraces many of the concepts of a ZCH. (Image courtesy Stewart Milne Group)

Load Factor = (actual kWh generated)/ (peak output kWh × 8760)

When comparing load factors over time it is important to remember that, for any given generation source, there will be large seasonal changes quarter-on-quarter and yearly variation may be significant. Different generating plants will operate at different load factors and these can vary from around 85 per cent for a Combined Cycle Gas Turbine (CCGT) power station, to around 30 per cent for a typical wind farm. In wind this is a particularly critical factor, since a small increase in average wind speed could result in a significant improvement in output and higher load factor.

PLANNING PERMISSION AND BUILDING CONTROL FOR RENEWABLES

Planning Permission

Planning permission requirements differ for each of the technologies. As a general guide, biomass, heat pumps and solar panels (thermal and PV) do not usually involve a planning application, whereas wind turbines and hydro plants do. The following general planning conditions, however, are worthy of note and it is always best to discuss project details with the local planning office at the earliest possible stage of its development. Planning permission applications will require a fee.

- Planning permission is normally required if the building associated with the renewable energy technology is a listed building, is in a Conservation Area, or is in an Area of Outstanding Natural Beauty (AONB), and the installation affects the external appearance of the building.
- Planning permission is normally required if you intend to construct a new building or build an extension to an existing dwelling to house the renewable energy technology, unless it falls within certain size and location parameters as laid down by the planning authorities.
- Planning requirements for community (non-domestic) buildings are different from the

requirements relating to renewable installations in domestic buildings, and will be assessed on a case-by-case basis; it is advisable to contact the planning authority before embarking on the project.

Building Control

Any changes made to a property, including installing a renewable technology, may require Building Control approval. Building Control applications are handled by the relevant local authority or council. Early contact with Building Control is recommended. An application fee is required and this is determined by the nature of the work being undertaken. In the UK a Building Control Certificate must be issued for all new heating systems including renewable energy heating. This includes wood fuel boilers and stoves, heat pumps, solar water heating, or indeed any modifications to existing heating systems. A Building Control Certificate is generally required for photovoltaic (PV) panels and building-mounted wind turbines. If approval is required, you will need to apply through your local Building Control Office. It would be advisable to contact the Building Control department for further advice on the application process. Unfortunately the issuing of Building Control Certificates will attract fees and experience has shown that these may vary between authorities. These certificates will be included with the Home Information Pack (HIP) that will be mandatory in the UK from 2008.

FINDING AN INSTALLER, MAKING IT HAPPEN

The most important aspect of installing any renewable energy technology is getting the right advice before the start and making sure it is carried through to completion. In most instances renewable energy equipment manufacturers, suppliers and installers will be different people or companies. It is essential that a full site evaluation and inspection is completed, that the most appropriate equipment is selected and that this is installed in line with the very best practice, using fully trained and accredited installers. Although the technologies have been around for a long time, the skills needed to ensure they are applied properly are still being developed.

Table 7: Typical power plant load factors

Technology	Typical Load Factor (%)
Small wind turbine	15–20
Wind farm	25–45
Hydroelectricity turbine	25–50
Biomass boiler	70–80
PV array	8–10
Combined Cycle Gas Turbine power station	70–85
Coal-fired power station	60–70

Load factors are typical and for illustration only. They can be much worse, and sometimes better, depending on the individual site conditions.

There are a number of agencies actively seeking to establish benchmarks for renewable energy installation standards and training. In the UK installers can only perform grant-assisted installations if they have been accredited under the UK Microgeneration Certification Scheme, managed on behalf of the Department for Business, Enterprise and Regulatory Reform (DBERR) by the Building Research Establishment (BRE). In Northern Ireland and the Republic of Ireland a new renewable energy training scheme has been developed, the Renewable Energy Installers Academy (REIA). This scheme has been developed using European INTERREG IIIA funding managed by Action Renewables and SEI (details of these organizations and relevant web links are given on pages 25–27). Installers who perform work under the Reconnect grants programme in Northern Ireland and the Greener Homes scheme in the Republic of Ireland must be registered with the REIA. World-class accredited courses have been developed across six technologies and installers are required to sign up to a customer protection Code of Practice.

As the individual technologies are reviewed in the following chapters, advice and guidance on their installation will be offered. Central to the success of any installation, both operationally and economically, is a good understanding of the installation and how it should be operated. A good installer will ensure that owners know their system, explain what the various bits are for and how they work, and will provide fully marked-up, as-installed drawings. This will help greatly when it comes to fault finding or when trying to work out which pipe or cable goes where. Adequate labelling and tidy work are essential components of any safe and competent installation, which should provide sustainable heat and electricity for many years ahead.

GLOSSARY

Building Control Regulations These exist to ensure the health and safety of people in and around buildings, and the energy efficiency of buildings. The Regulations apply to most new buildings and many alterations of existing buildings, whether domestic, commercial or industrial.

Carbon footprint An indication of the carbon dioxide emitted as we carry out our daily lives. It covers all aspects of our energy use, including transportation (including holidays), food consumption and waste disposal.

Embedded generation Generators (not necessarily renewable) located at premises that are not large-scale generation sites (power stations). These generators bring a number of benefits to both the electricity grid system and the customer through reduced losses and network relief at weaker areas on the electricity system.

Kelvin The SI unit of temperature, expressed in degrees Kelvin.

Low-grade heat Usually regarded as heat that is not of a high enough temperature to directly satisfy the space and water heating requirements of buildings.

Ofgem Office of Gas and Electricity Markets (England, Wales and Scotland).

Permafrost layer Layer of soil at or below the freezing point of water, located in high latitudes (i.e. in the geographical regions surrounding the North and South Poles) or at high altitudes.

Positive-feedback As the result of certain actions, there will be further similar, reinforcing actions. In this instance, increased global temperature will result in releases that cause further temperature increase, which in turn will cause greater releases, increasing temperatures further, etc.

Sell and Buy-back Agreement A form of contract between a renewable generator and an energy supply company, whereby the energy company will purchase the total output from a renewable generator and sell the on-premises consumption back (as a licensed supplier).

White Paper Informal name for a parliamentary paper setting out government policy or proposed action on a topic of current concern.

USEFUL RENEWABLE ENERGY CONTACTS AND WEBSITES

AGORES

www.agores.org
A European Union site with extensive links to organizations and manufacturers all over the world.

CADDET
www.caddet.org
Specializes in documenting demonstration projects on renewable energy and energy efficiency right across Europe and the world. CADDET's database of equipment suppliers, GREENTIE, supplies information on approved renewable energy appliances and equipment, including those available in the UK and Ireland.

The Centre for Alternative Energy (CAT Centre)
www.cat.org.uk
Practical information on most aspects of renewable energy.

CREST
www.lboro.ac.uk/departments/el/research/crest/
Based at Loughborough University, CREST carries out research into renewable energy.

Department for Business, Enterprise & Regulatory Reform (DBERR)
www.berr.gov.uk/energy/index.html
DBERR (previously DTI) is the ministry responsible for UK energy policy. The site contains many renewable energy policy documents and statistics. The 'Renewables Explained' page gives access to further pages about renewable energy technologies and projects, including many outside the UK. The DBERR site is a good source of research reports and papers and it carries a description of the various renewable technologies.

The Open University
www.open.ac.uk/T206/Weblinks/intro.htm
Associated with the OU's renewable energy course T206, this site links to many of the most useful and popular renewable energy sites worldwide.

Renewable Energy Association
www.r-p-a.org.uk/home.fcm
Previously known as the Renewable Power Association, REA is a trade association representing the various technologies.

Reports

Renewables Innovation Review
www.berr.gov.uk/energy/sources/renewables/publications/page26280.html
Survey undertaken by DBERR in 2005 to review the state of the UK renewable energy technologies, including barriers and opportunities.

General Grants Information

Low Carbon Building Programme (LCBP UK)
www.lowcarbonbuildings.org.uk/home
www.energysavingtrust.org.uk/housingbuildings/funding/lowcarbonbuildings/
A UK grants programme (including PV) managed on behalf of the Department for Business, Enterprise and Regulatory Reform (DBERR) by the Energy Saving Trust.

Grants and Information in Scotland

Scottish Community and Householder Renewables Initiative (SCHRI)
Tel: 0800 138 8858
www.est.org.uk/schri
A Scottish Executive-funded scheme offering grants, advice and project support to develop and manage renewable energy projects, including pellet heating systems, for both households and communities (no PV grants in Scotland).

Grants and Information in Northern Ireland

Reconnect Household Grant
www.reconnect.org.uk
Funded through DETI (NI), including PV. Programme managed by Action Renewables (www.actionrenewables.org).
Freephone 0800 0234077

continued overleaf

USEFUL RENEWABLE ENERGY CONTACTS AND WEBSITES

Solar PV Grants

UK contact: **Energy Saving Trust (EST)**
www.lowcarbonbuildings.org.uk/home
21 Dartmouth Street, London SW1H 9BP.
Tel: 020 7222 0101 Fax: 020 7654 2460
Details of UK government grants for systems under the Major Photovoltaics (PV) Demonstration Programme.
www.energysavingtrust.org.uk/housingbuildings/funding/solarpv/

Northern Ireland Electricity (NIE)
www.NIEyourenergy.co.uk
Tel. 028 90 661100
Offers support grants for PV panels for domestic, schools and industry in Northern Ireland.

Renewable Energy Information and Grants in the Republic of Ireland

Sustainable Energy Ireland (SEI)
www.sei.ie/greenerhomes/
Greener Homes is the Republic of Ireland's domestic grants programme.

Renewable Energy Installers
UK Microgeneration Certification Scheme
www.ukmicrogeneration.org
Email: microgeninstallers@bre.co.uk
A scheme operating in the UK and managed on behalf of the DBERR by Building Research Establishment (BRE) Certification.

Renewable Energy Installer Academy (REIA)
www.reinstalleracademy.org
Maintains lists of suitably accredited installers in Northern Ireland and Republic of Ireland.

Utility Regulators

Ofgem – England and Wales
www.ofgem.gov.uk
9 Millbank, London SW1P 3GE
Tel: 020 7901 7000 Fax: 020 7901 7066

Ofgem – Scotland
www.ofgem.gov.uk
Regent Court, 70 West Regent Street, Glasgow G2 2QZ
Tel: 0141 331 2678 Fax: 0141 331 2777

Northern Ireland Utility Regulator (NIAUR)
www.niaur.gov.uk
Queens House, 14 Queen Street, Belfast BT1 6ER
Tel: 028 9031 1575 Fax: 028 9031 1740

Commission for Energy Regulation (CER)
www.cer.ie
The Exchange, Belgrade Square North, Tallaght, Dublin 24
Tel: 00 353 (0)1 4000 800 Fax: 00 353 (0)1 4000 850

Green Electricity Suppliers

www.ecotricity.co.uk
www.goodenergy.co.uk
www.greenenergy.uk.com
www.greenelectricity.co.uk
A supplier comparison site is available at www.whichgreen.com.

Other Useful Contacts

The Environment Agency
www.environment-agency.gov.uk
Rio House, Waterside Drive, Aztec West, Almondsbury, Bristol BS32 4UD
Tel: 0845 9333111 Fax: 01454 624 409

Community Renewables Initiative
www.cri.energyprojects.net
Email: communityrenewables@esd.co.uk
The Countryside Agency, John Dower House, Crescent Place, Cheltenham GL50 3RA
Tel: 01242 521 381 Fax. 01242 584 270
Helps local people and organizations devise and implement renewable energy developments. Local support teams around the country.

USEFUL RENEWABLE ENERGY CONTACTS AND WEBSITES

Health and Safety Executive (HSE)
www.hse.gov.uk
Rose Court, 2 Southwark Bridge, London
SE1 9HS
Tel: 020 7717 6000 Fax: 020 7717 6717
For Health and Safety regulations, contact the HSE InfoLine:
Tel: 0870 545500 Fax: 02920 859260
Email: hseinformationservices@natbrit.com

The Stationery Office (formerly HMSO)
www.clicktso.com
Tel: 0870 600 5522 Fax: 0870 600 5533
Publishes national standards, health and safety regulations and codes of practice.

Planning Issues (GB mainland)
www.planningportal.gov.uk
Approved Document J: Combustion Appliances and Fuel Storage Systems, available at www.planningportal.gov.uk/england/professionals/en/4000000000503.html
Approved Document G3: Hot Water Storage, available at planningportal.gov.uk/england/professionals/en/4000000000371.html

Scottish Building Regulations
www.scotland.gov.uk/buildregs
Part F is the equivalent of Part J in England and Wales.

Standards and Technical Regulations Directorate
www.safed.co.uk/Regulators/index.htm
For information on the Pressurised Equipment Directive Guidance notes on the UK Regulations, Nov 99; URN 99/1147.

Water Regulations relevant to wet heating systems.
www.wras.co.uk

Pressure Systems Safety Regulations
www.hse.gov.uk/pubns/psindex.htm
Guidance for operators and owners.

Renewables East
www.renewableseast.org.uk
In-depth information about renewable energy, with particular reference to eastern England.

Renewables North West
www.renewablesnorthwest.co.uk
In-depth information about renewable energy, with particular reference to north-western England.

Regen South West
www.regensw.co.uk
In-depth information about renewable energy, with particular reference to south-west England.

Rerum
www.rerum.org
A programme for municipalities to realize renewable energy projects, as well as rationalizing their use of energy.

Scottish Renewables Forum
www.scottishrenewables.com
Scotland's Forum for the Renewable Energy Industry.

South East England Renewable Energy Statistics
www.SEE-Stats.org
The definitive source of statistics and listings for all renewable energy sources in South East England.

TV Energy
www.tvenergy.org
Promoting renewable energy in the Thames Valley, Surrey and northern Hampshire.

CHAPTER 2

SOLAR WATER HEATING

SWH panels will become 'must have' items in every existing or new home, as is now the 'norm' with double glazed windows. They should be fitted as mandatory to every new building with a south-facing roof, so that the occupants can enjoy around 40 to 50 per cent of their hot water, for free, over the lifetime of the panel, indeed, potentially, the lifetime of the building.

HOT WATER FROM SUNSHINE

Contrary to popular opinion, the British Isles does have a useful resource of solar energy and this can be used to reduce the space heating demand of new buildings through passive solar design. The Sun can also be used to heat air and water, through active solar heating systems, and it can provide electricity through the use of photovoltaic systems (*see* Chapter 6). A group of solar panels is known as an array.

The main components of any solar water heating (SWH) system are the solar collector(s) or solar panels, a heat transfer system to move the heat from the collector to the point of use (known as the heat transfer loop or primary circuit), and an insulated reservoir or tank that stores the heat for subsequent use (the hot water cylinder). The heat transfer fluid most commonly used in the primary circuit is a mixture of water and glycol antifreeze, sometimes referred to as brine.

Solar panels work best when they are mounted on south-facing roofs, but they can work just as well on the south-facing side of a building or on ground-mounted support frames – anywhere in fact with an unobstructed view of the Sun over the longest period of every day. These systems provide domestic hot water, occasionally heat swimming pools and, much less frequently, they can provide space heating. They are particularly well suited for hotels and guesthouses and are ideal for the agricultural community for use in dairy farms. Solar heat energy can also be used widely in industrial applications and it can provide the necessary energy input for other uses, such as cooling equipment. A SWH system operating in the British Isles can produce 40 to 50 per cent of the annual hot water requirement for an average family, the vast majority becoming available over the months from May through to September. The use of solar energy to provide space heating will only be described briefly, since it is rarely an economic proposition in the latitudes of the British Isles and will require a back-up source of heating during the winter months.

SWH systems are suitable for use with most conventional heating systems (coal, oil and gas-fired central heating) and they are becoming increasingly paired with biomass-fuelled systems (*see* Chapter 4). The exception to this is the combi-boiler, which does not have a storage tank, although a few more recent

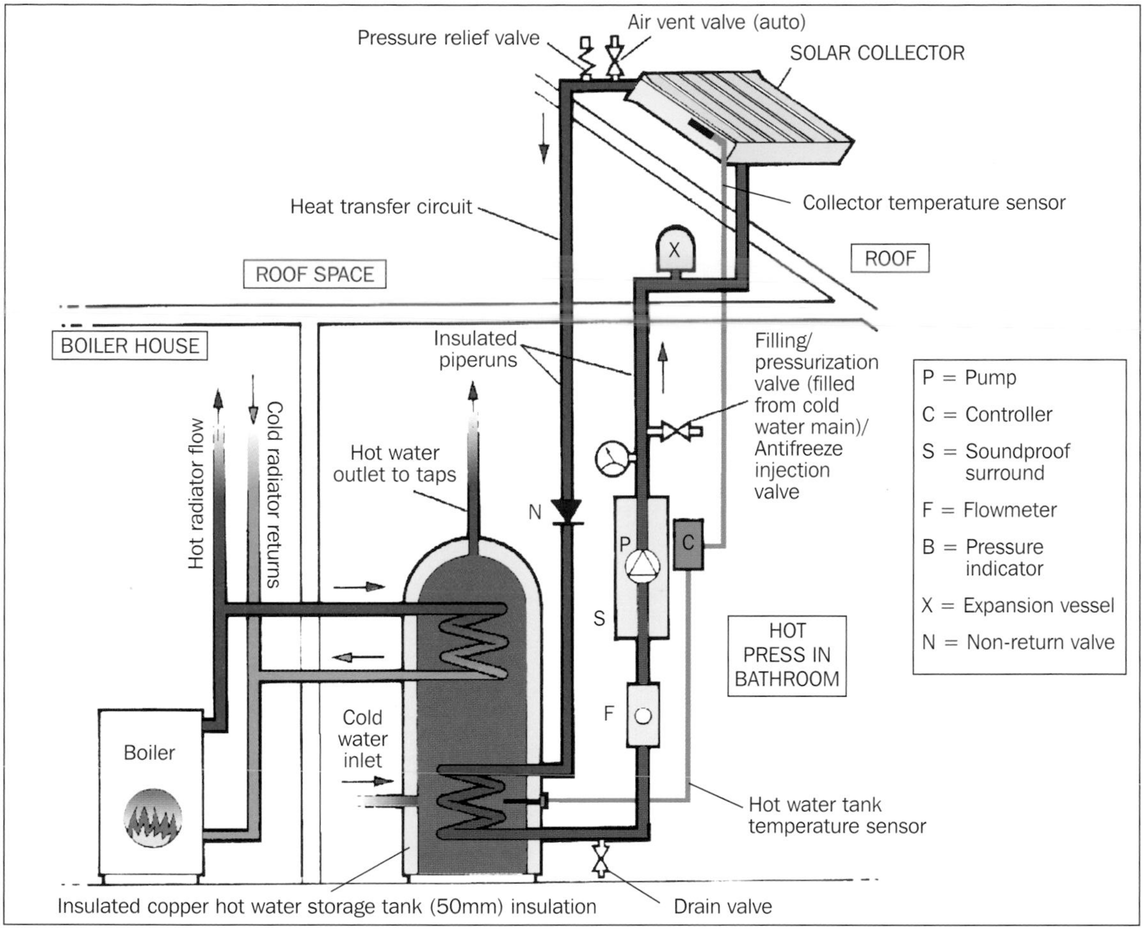

Pressure relief valve
Air vent valve (auto)
SOLAR COLLECTOR
Heat transfer circuit
Collector temperature sensor
ROOF SPACE
ROOF
BOILER HOUSE
Insulated piperuns
Filling/ pressurization valve (filled from cold water main)/ Antifreeze injection valve
Hot radiator flow
Cold radiator returns
Hot water outlet to taps
N
X
P
C
S
F
HOT PRESS IN BATHROOM
Boiler
Cold water inlet
Hot water tank temperature sensor
Insulated copper hot water storage tank (50mm) insulation
Drain valve

P = Pump
C = Controller
S = Soundproof surround
F = Flowmeter
B = Pressure indicator
X = Expansion vessel
N = Non-return valve

Figure 7: Main components of a SWH system.

combi-boiler designs can accept hot water input from a solar-heated, pre-heater feed tank.

SWH systems are usually distinguished on the basis of their solar collector design, either 'flat plate' or 'evacuated tube'; they may also be either direct or indirect systems. Most systems are simple to install, although health and safety concerns are very important, particularly because of the need to work on roofs and due to the high water temperatures (around 180°C) that might be reached under some operating conditions. In summer these systems can produce 100 per cent of an average family's hot water requirement, but clearly this cannot be the case in the winter. Over the winter months only 10 to 15 per cent of the hot water demand can be expected.

Figure 8: Three Schuco flat plate solar panels fitted to a domestic house. (Photograph copyright Mark Compston, 2007)

Heat Transfer Mechanisms

Heat may be transferred in three ways and these are all employed in one form or another in a SWH system. **Thermal energy** (heat energy) moves from one place to another because of a **difference in temperature** between them. The energy transfer is always from **hot** to **cold.** In conduction, energy is passed from atom to atom. It is the most important type of heat transfer in solids such as metals. In the collector of a SWH panel, heat energy is transferred from the absorber (which gets hot) to the fluid in the collector (which is cooler) by conduction.

In convection, energy is carried around by atoms or molecules, so it only happens in liquids and gases. In the solar panel, heat will be lost by convection from the absorber through contact with the surrounding air. In evacuated tube collectors, the air in the tube is removed (hence the name) and the heat loss by convection is minimized. The third heat transfer mechanism is radiation: radiant energy (at infrared wavelengths) is given off by a hot object (the Sun, for example) and is absorbed by another object (the absorber on the solar panel). The object of the solar collector is to retain as much of the incident radiation as possible by not re-radiating it, to minimize heat loss by conduction (through solar panel insulation) and to prevent convective losses so that the maximum amount of heat can be transferred to the heat transfer fluid in the primary circuit.

THE DEVELOPMENT OF SOLAR WATER HEATING

The earliest SWH systems consisted of little more than water tanks painted black and placed behind sheets of glass. The first examples of these systems appeared on the frontier homes of the USA. The Climax was an early commercial system marketed around 1890 by Clarence Kemp, a heating salesman from Baltimore. The modern solar water heater was first patented in 1909 by William J. Bailey in California. It was similar in design to the thermosyphon system (see below). Bailey discovered that a layer of insulation placed around the water storage tank could keep the water hot overnight and he called his company The Day and Night Solar Water Heater Company. The concept boomed during the early decades of the twentieth century; 80 per cent of new homes built in Miami, Florida, between 1935 and 1941 had SWH fitted. It is estimated that during that period as many as 60,000 systems were sold in that area alone. The demise of the nascent industry came with the discovery of cheap gas and oil deposits.

The oil price scares of the early 1970s stimulated the rebirth of SWH systems with major markets appearing in Spain, Germany, Japan and many of the Middle Eastern and Asian countries. SWH technology has steadily improved with the use of insulation and the introduction of evacuated tubes, which prevent almost all of the collected solar heat from escaping. Today, in the first decade of the twenty-first century, the looming threat from climate change, the presence of government grants and a sharp increase in fossil fuel prices right across Europe, particularly in the UK and Ireland, has prompted a rise in interest in these systems.

SOLAR WATER HEATING IN THE HOME

Domestic SWH, providing hot water for baths, showers, washing dishes and other uses, represents the best application for active solar heating in our climate, which is very useful since this accounts for around 6 per cent of our total energy use. Domestic hot water is required throughout the year, including the summer period, when there is an abundance of solar energy. Hot water tanks store water at around 60°C and a typical home uses approximately 5kWh of heat per day in the form of hot water (roughly 1,800–2,000kWh per annum). It is important to remember that the amount of hot water consumed can vary significantly from house to house depending on occupancy levels, the age of occupants and many other specific requirements for hot water. Running a fossil-fuelled central heating system over the summer months to provide hot water can be very inefficient, since the boiler will run for short periods, wasting fuel and never reaching full operating temperature, and this will have a detrimental effect on its lifespan. It is common practice to use electric immersion heaters to provide hot water in summer, though again this is both expensive and an inefficient use of the primary energy (electricity).

The United Kingdom's Department of Business, Enterprise and Regulatory Review (DBERR, formerly Department of Trade and Industry, DTI) completed a survey of 700 SWH systems fitted between 1970 and 1994. The sample included a range of collector designs and system sizes. The survey concluded that SWH systems can work reliably for twenty to twenty-five years and beyond, and advances in recent years have produced significantly improved designs. Half the systems required no new components during the period of survey and only 1 per cent were considered unreliable. Newer systems have fewer problems, indeed the survey showed that 68 per cent of the systems installed since 1986 required no new components. Developments over the last ten years or so have produced modern systems that are both highly efficient and very reliable.

Solar Powered Space Heating

The main requirement for space heating a building is at night, when there is no sunlight, and during the winter, when the amount of energy received from the Sun is lower. In the British Isles the solar gain (the energy received from the Sun) in winter is regarded as insufficient to provide adequate heating for an average home without support from a secondary heating system, such as coal, gas or oil-fired heating back-up. This does not mean that solar heating systems cannot be used for space heating, indeed it is estimated there are more than 3,000 such systems in operation in the UK.

A properly designed system can produce space heating from April to October. To provide a meaningful contribution, the system will require a large collector area and this infers that any available south-facing roof will need to be covered with solar collectors. Even an extremely large collector area will not provide any hot water on a dark, cloudy winter's day. The effectiveness of solar-powered space heating can be increased by using under-floor heating, especially since the thermal mass of the floor can act as a buffer store for heat (in the same manner as an electric storage radiator). The heat flow to the floor is dependent on the available solar radiation, so the back-up heating system will need to be able to increase or reduce its output to compensate (this is referred to as modulation).

Agricultural and Commercial Applications for Solar Water Heating

There are a number of applications in this sector that require large quantities of hot water on an almost daily basis. These systems are essentially scaled-up domestic installations.

Solar Heated Swimming Pools

Some estimates suggest that there are around 100,000 swimming pools in the British Isles, with roughly 70 per cent of them outdoors in private gardens. The majority are not used in the winter months. The control of water quality and temperature in swimming pools is complex and some pools are fitted with heat recovery systems and heat pumps (*see* Chapter 3). Not all of these pools will be suitable for solar heating, especially since large areas of solar collector will be necessary. A direct solar water heating design is normally the most appropriate for swimming pools (see below). A more recent application for SWH, of increasing interest in the United Kingdom, is the heated outdoor hot tub or jacuzzi.

As a rule of thumb, the area of the SWH collector panel should be 25 per cent of the surface area of the swimming pool. In some arrangements, the solar circulation loop heat exchanger is in the pool. If this is the case, the heat exchanger is manufactured from stainless steel to prevent corrosion due to the chlorine in the pool water.

DESIGNING A SOLAR WATER HEATING SYSTEM

Direct and Indirect SWH Systems

There are two SWH configurations: direct and, by far the more common, indirect systems. In a direct system, regardless of collector type, the water flowing in the collector tubes of the solar panel is the same water that carries the heat to, and circulates in, the hot water storage tank, and it is also the same water that comes out of the hot taps. Direct systems have

Solar Radiation and Solar Panel Orientation

We see solar radiation as white light, but in fact it is composed of a spectrum of wavelengths from infrared (longer wavelength than red) to ultraviolet (shorter wavelength than blue). Solar collectors depend on the property of glass that permits full-spectrum solar radiation to pass through, but blocks this radiation when it is re-radiated as longer-wave infrared or heat radiation; this process is in fact the 'greenhouse effect'. When radiation from the Sun strikes the Earth's atmosphere, it either passes straight through (as beam radiation) or it is scattered and becomes diffuse radiation (or diffuse lighting). Direct radiation is the sunlight that appears to come directly from the Sun; it is surrounded by a region of less intense sunshine or diffuse radiation. On a cloudless, sunny day direct radiation can approach a power density of 1 kilowatt per square metre (1kW/m^2), or '1 sun' for solar collector test purposes. In northern Europe this figure is closer to 900W/m^2 at ground level. At mid-northerly latitudes, it is assumed that around 50 per cent of the annual sunshine is collected as direct sunshine and the balance is as diffuse radiation. Only the direct radiation can be focused in solar concentrators, whereas it is the diffuse radiation that provides most of the internal daylight in buildings, for example in rooms with north-facing windows or rooms that do not have a direct solar exposure. Surfaces that absorb and emit heat are referred to as 'black bodies' and the absorbers in solar collectors perform as black bodies.

The solar energy resource can be assessed using devices known as solarimeters. These measure the total solar energy that falls on horizontal surfaces (also called the insolation). More detailed measurements can determine the diffuse and direct components of the solar radiation and these can be used to estimate the energy incident on tilted or vertical surfaces. There is considerable variation in the radiation levels, and hence the collected solar energy between summer and winter in the British Isles. For example, in mid-July the

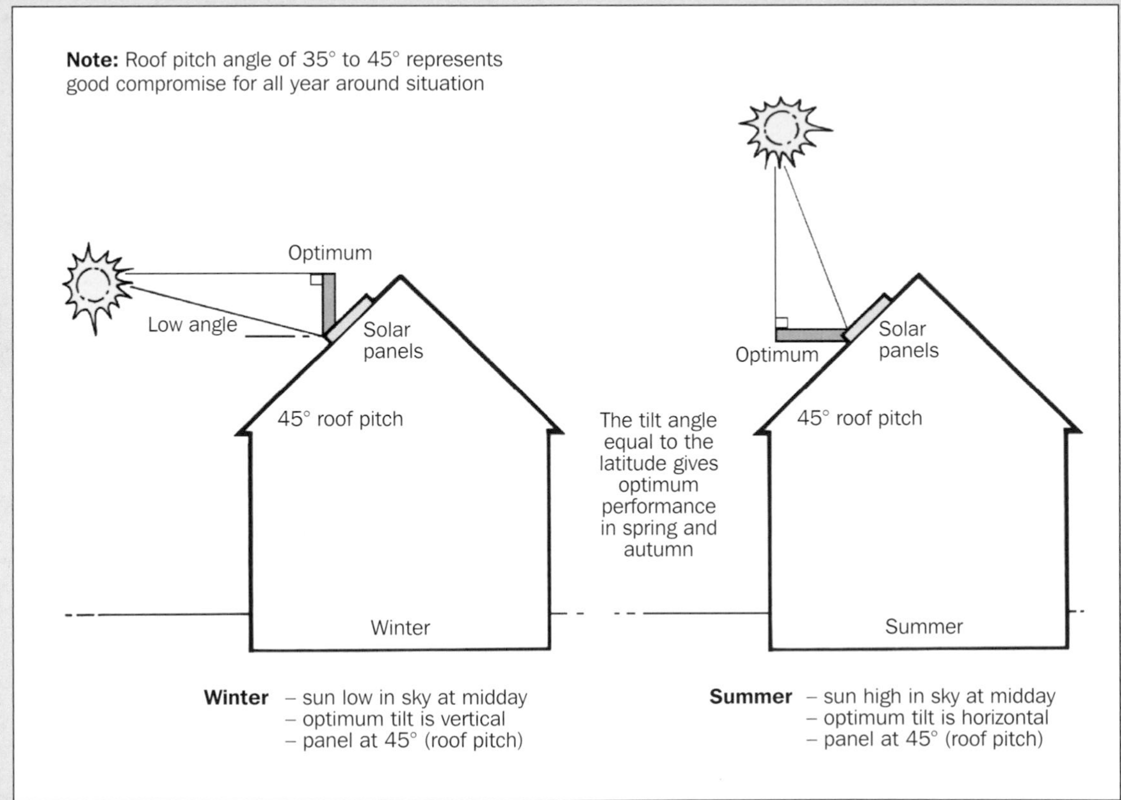

Figure 9: The tilt angle of the solar panel may be altered to optimize the solar energy capture.

solar energy falling on a horizontal surface over the period of a day is around 4.5kWh/m^2. In January this figure falls to around one-tenth of its summer value, about 0.5kWh/m^2. The implications are that the solar resource is more useful in summer and, although there is a resource in winter, it is less intense and more difficult to collect. In periods of cold weather with chilling winds the effectiveness of solar collectors is further reduced.

Solar collectors should be fastened to the roof of a building in a position that will receive as much sunshine as possible. The best orientation is facing the Sun at midday, which in the British Isles is south and tilted at an angle. This tilt angle is dependent on the latitude and the time of year. If the collector is tilted at exactly the latitude angle it will be perpendicular to the Sun's rays at midday in March and September. The tilt should be a little more horizontal to maximize summer collection and tilted towards the vertical for optimum winter collection. The tilt angle is not critical and, for most applications, panels mounted at the latitude angle will provide a good compromise. For northern latitudes, where solar heating systems will provide the best output in the summer months, it is recommended that the panel should be tilted at the latitude angle minus ten degrees (around thirty to forty-five degrees for the UK and Ireland); this coincides with the normal slope angle for pitched roofs. If a tilted roof is not available or the orientation is outside the desired range, then wall- or ground-mounting on a specially designed angled bracket remains an alternative. Flat roofs, if they are sufficiently load-bearing, can also be used to support a frame-mounted panel or array.

Variation in panel orientation away from the optimum (due south) will result in reduced output, although again this is not significant if the displacement is small. Practical constraints may mean that collectors face in a direction between south-east or south-west, but this will not result in a serious degradation in performance. A large proportion of existing buildings will have a southerly aspect and would be suitable for the installation of SWH systems. Orientations outside this, however, for example collectors facing due west, east or even north, will be cost-inefficient and should be avoided.

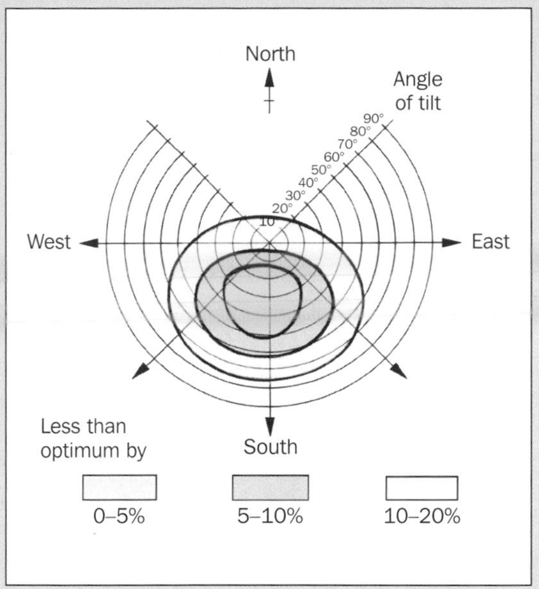

Figure 11: Influence of orientation and tilt on solar panel output. (Reproduced courtesy of the Renewable Energy Installer Academy [Action Renewables & SEI]).

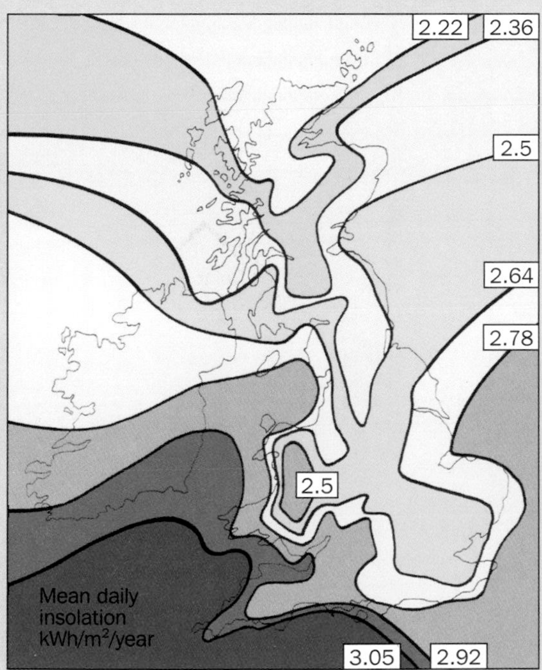

Figure 10: Approximate annual average levels of solar radiation for the United Kingdom and Ireland. (Adapted from McNicholl and Lewis and reproduced courtesy of Patrick Waterfield)

the advantage that an additional storage tank or a new tank with an additional coil is not required. In some retrofit situations, especially where space is tight or minimum disruption is essential, it might be preferable to retain the existing tank and in these cases a direct system may be more appropriate. Direct systems present more of a problem in situations where freezing can occur, since it is impossible to put antifreeze in the water. Direct systems can have better efficiencies than indirect systems, but they might have higher running costs in hard water areas. If a system is installed in a hard water area, then it will have a higher maintenance requirement due to corrosion and the need to descale furred-up pipes. Drain back or drain down systems (see below) are used to prevent freezing with direct-connected SWH by allowing the water in the collector to drain back into the storage tank.

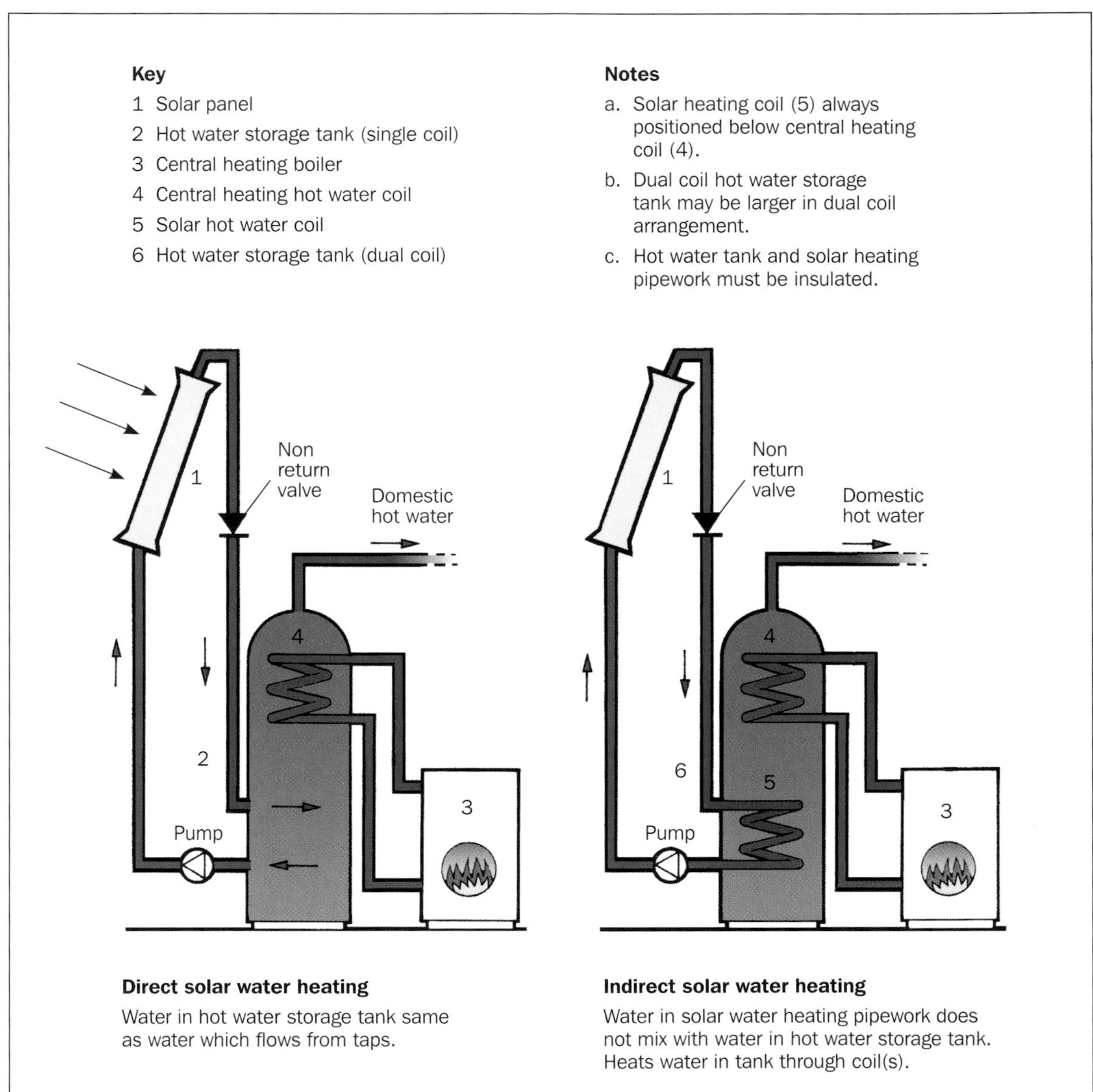

Figure 12: Direct and indirect SWH showing main components of each system.

Most systems currently being installed are indirect systems and they will usually include a replacement hot water storage tank, which is larger than the original and must be well insulated. In an indirect system the heat transfer fluid in the collector does not come in direct contact with the water in the hot tank. Instead a heat transfer fluid flows in the collector and in the pipes to and from the collector to the hot tank, known as the primary circuit. Apart from avoiding the hazards from freezing, indirect systems have the advantage that the SWH system primary circuit fluid, with its contaminants due to corrosion, additives and scaling, never comes into contact with potable water. Small bore pipe runs can be used between the panels and the tank, and the sealed primary circuit means that furring in hard water areas is not a problem. Most indirect solar water systems use a primary circuit that is pressurized to improve efficiency of heat transfer and this will require the addition of a pressure expansion vessel.

Drain Back or Drain Down SWH Systems

To avoid the problems associated with freezing, some systems allow the water in the collector primary circuit to drain down back into a drain back vessel or, in some cases, back into the hot water storage tank. In this arrangement, the heat transfer fluid in the solar panel is circulated as normal through the coil in the hot water tank by a pump. However, when the pump is switched off, valves operate to allow the water to drain back into the drain back vessel leaving the collector pipework and pipe runs empty. This configuration has the advantage that there is no water in the primary circuit to freeze and no antifreeze solutions are necessary. The circulation of the heat transfer fluid in the primary circuit can also be stopped when the temperature in the collector goes too high. In this situation the water simply drains back into the drain back vessel. This is good protection against a system overheating in very sunny weather when there is little or no demand for hot water, for example when the building's occupants are on holiday. The downside is that the drain back system will switch off during bright, cold periods when the air temperature is below freezing, but there may still be enough solar radiation to heat the water. A drain back system will also switch off during very hot periods when pressurized systems can keep working. Both flat plate and evacuated tube collectors can operate with drain back systems, but unless a further pre-heat tank is used these systems will be direct.

Active and Passive SWH Systems

Active SWH involves the use of a pump to actively circulate the heat transfer fluid through the primary circuit, that is, to draw cool water from the bottom of the hot water storage tank (also called the hot water cylinder) and pump it to the solar collector, and then to return the heated water from there back to the hot water storage tank. Passive systems do not have a pump. Instead they rely on natural circulation, which occurs when hot water rises and cooler water sinks to take its place, forming a circulation within a closed system. This natural (unpumped) circulation is called thermosyphonic circulation.

Systems with a pump will require a pump controller. Control systems for SWH are straightforward. The temperature of the heat transfer fluid in the solar collector is constantly compared with the water temperature in the hot water storage tank. When there is sufficient differential between the two (the collector temperature is around five to eight degrees hotter than the storage tank), the pump will switch on automatically, pumping the hot water from the collector on the roof into the tank. The controller monitors this differential and the pump will run until the differential falls below a preset value (usually 2°C).

An example of a passive SWH system used extensively in the sunnier regions of the world is the thermosyphon SWH system. This is a direct system and consists of a solar collector, a storage tank for the heated water and the associated connecting pipes. Very often the whole assembly is mounted in a modular frame. The storage tank must be sited sufficiently above the collector to allow natural circulation to take place. In modern domestic properties the positioning of the collector can be restricted. Although cheaper and more reliable than the other systems described, thermosyphon systems tend to be less efficient in diffuse sunlight conditions. A typical system for a four-person home consists of a tank of 150 to 300 litres and three to four square metres of flat plate solar collector panels. In sunnier climates these systems are very common, with solar collectors and storage tanks mounted on hotels, garages and shops.

Auxiliary Heating Sources

SWH systems are normally fitted to hot water systems that already include other heating sources. These auxiliary sources could be an electric immersion heater and/or a coil in the storage tank, which is heated indirectly by a central heating boiler. Hot water can therefore always be available, even with the prolonged absence of solar energy, and the danger from bacterial growth in the water (Legionella or Legionnaire's Disease) is minimized. The temperature of the water in the hot water storage tank must be raised above 60°C at least once every day to prevent the growth of these bacteria.

Pre-Heat Storage Tanks

Sometimes, when a central heating system is used to heat the water in the hot water storage tank, the hot water from the solar heating system will be held in a pre-heating tank. This tank will act as a heat store and the feed water for the main hot water storage tank can be pre-heated by passing it through a heat exchange coil in the pre-heater tank. Solar pre-heater storage tanks can be used in situations where space is restricted, or where it is impossible to access or move the existing hot water storage tank. As mentioned above, certain combi-boiler systems can also draw their input supply from pre-heater tanks storing the solar hot water. When the solar water heat exchanger (the solar coil) is fitted to a main hot water cylinder, it must be located below the main heating system heat exchanger coil.

It is important to remember that heat will circulate naturally in a SWH system and this means that a solar panel may also radiate heat from the storage tank at night, or at times when the tank temperature is higher than the temperature in the collector on the roof. To prevent this, a non-return valve (or check valve) is fitted in the primary circuit to ensure hot water flow is in one direction only, from the collector to the tank. It is also important that all pipework around the hot water storage tank is well insulated to prevent heat loss, including the hot water draw-off pipework for the building.

SOLAR PANEL COMPONENTS

Solar Collectors

The solar collector is the principal component in any SWH system. Its job is to collect the solar radiation and transfer it to the heat transfer fluid to raise the temperature of the water in the hot water storage tank. Approved collectors are required to meet a European standard, BS EN 12975:2000 'Thermal Solar Systems and Components – Solar Collectors'. This standard specifies requirements for, and sets down a means of measurement of, the durability, mechanical strength, reliability and safety of solar collectors, as well as specifying the details of a recommended performance test. Two designs of solar collector are most commonly found, the flat plate collector and the evacuated tube collector.

It is generally considered that the evacuated tube designs are more efficient. This is reflected in the lower surface area needed for the panels, and hence smaller physical size, although they also come at a higher cost. Both types of collector blend well with most roof appearances, resembling skylight windows when viewed from the ground.

As a general rule, around $0.7m^2$ of collector surface per occupant can be used to estimate the required collector area for evacuated tube systems and $1.0m^2$ per occupant can be used for flat plate systems.

SWHs are designed to satisfy the expected demand for heat take-off to the hot water system. The absorber coating will decide how hot the surface becomes and systems that demand higher or constant supplies of hot water should be designed to have absorber coatings that will reach higher temperatures.

At times the fluid in the primary circuit, and ultimately the hot water storage tank, can reach very high temperatures, especially during prolonged periods of strong sunshine coincident with a low or non-existent demand for hot water, or if the pump fails. Designs should allow for relief of pressure and excess heat through a pressure relief valve and heat dump when these operating conditions are experienced. It is essential that relief valves are piped to safe release areas, such as outdoors at ground level. To avoid the risk of serious scalding, relief valves and vents must not release hot water or vapour at head-height indoors, or inside hot presses and cupboards.

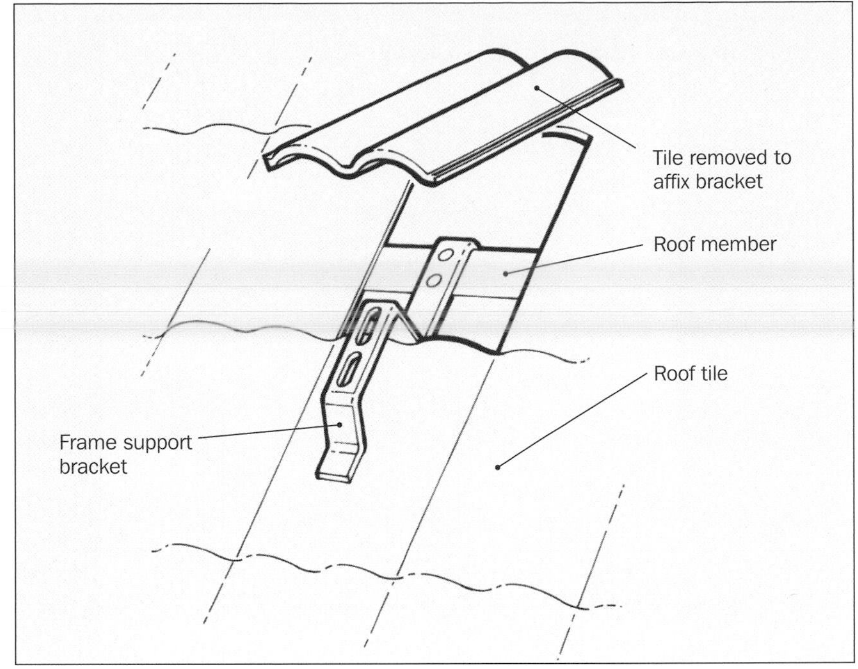

Figure 13: Solar panel fixing frame mounting bracket (detail). The mounting bracket is fixed securely to the roof timbers. Tiles may be slid upwards for fixing and returned to their original position once the bracket is in place.

Flat Plate Collectors

These are the most cost-effective and widely available solar panels. They are normally constructed as a rectangular box with a single- or sometimes double-glazed top panel to admit the solar radiation. The more glazing the better the light absorption, although around 7 to 10 per cent of the incident light is lost through reflection, from the glazing and the absorber. Plastic sheets rather than glass may be used to afford greater protection against impact damage (as a result of rough handling or vandalism, for example), although the plastic needs to be transparent to absorb solar radiation. The panel must contain an absorber plate from which heat can be removed by a heat transfer fluid. The absorber plate is normally painted matt black (or dark navy blue) for high absorptivity, although this may still reflect around 10 per cent of the incident radiation. Surfaces that have high absorptivity in the visible region, with low emissivity in the long-wavelength infrared, will reduce heat losses.

Collector casings are usually made from pressed aluminium and absorbers are small bore copper pipes, soldered to thick copper or steel sheets. It is essential that the absorber has a high thermal conductivity to transfer the collected heat to the working fluid in the pipes with minimum temperature loss. The collector needs to be insulated using mineral wool or foam insulation at the back and sides of the absorber plate and the whole assembly contained in a weatherproof casing. Weight presents a relative disadvantage for flat plate systems, as the panels have to be raised up onto roofs and the flat plate's size and weight are greater than that of an evacuated tube panel. More recent flat plate designs are lighter through the greater use of plastic casings to facilitate easier handling. Flat plate collectors are more resistant to impacts and mechanical shock.

Flat plate solar panels are attached to specially designed metal frames, which are fitted on top of the tiles or slates of the existing roof covering. Brackets are then slipped below the roof covering and attached directly to the roof structure by screwing them to the existing lathes and roof timbers. The brackets support the framework to which the panels are attached.

There has been increasing demand, especially in new build, for roof-integrated solar collectors to be built into the roof during construction. Flat plate collectors are especially suited to this and factory made in-roof panels are available. The panel must be modified to accept roof flashing kits for full

Flat Plate Collector Components

A collector cover must have good resistance to bad weather and to atmospheric pollutants, tolerate wide temperature variations and have the material strength to withstand snow and wind loading. The cover also requires high transmission of visible and near infrared radiation (wavelengths up to 2.4 micrometres) to maximize the radiation incident on the absorber and minimum transmission of infrared (wavelengths greater than 3 micrometres) to prevent the heat energy re-radiated by the absorber escaping back to the atmosphere.

Glass is most commonly used, although British float glass has a transmittance of only 86 per cent and is heavy. Blow-formed 3mm acrylic is strong and lighter than glass and has a transmittance of 89 per cent. Other options are double-walled 10mm polycarbonate plate and plastic films, such as Teflon or Tedlar, which have high resistance to ultraviolet radiation. These materials can sag with heat, sometimes attract dirt and are lower in strength, but they are lightweight, have high transmittance and are relatively cheap.

Insulation is used in the solar panel to limit the heat loss to the surroundings. It must be capable of withstanding high temperatures (over 150°C); it must not melt nor outgas, even at these elevated temperatures, and it must be free of chlorofluorocarbon (CFC) materials. Mineral wool is frequently used in flat plate collectors.

Collector casings are made from glass reinforced plastic (GRP), aluminium or stainless steel. The box corners are sealed to prevent the ingress of rain and they are ventilated to avoid condensation. The vent holes also make sure that excessive air pressure does not build up in the panel when the components expand due to high temperatures.

weatherproofing. Less expensive unglazed plastic collectors are available, but these tend to be used exclusively for outdoor swimming pools, where lower water temperatures are acceptable.

Evacuated Tube Collectors

A typical evacuated tube collector design, as illustrated in Figure 16, has parallel rows of long, glass cylinders. Evacuated tube collectors are claimed to be more efficient than flat plate collectors, performing well in conditions of direct and indirect sunshine. They are not used in direct SWH systems. In some evacuated tube systems, tubes may be added or removed to customize and accommodate changes in hot water requirements. The absorber plate is a metal strip or fin running down the centre of each tube. Losses through conduction and convection are reduced by the vacuum (or partial vacuum) that exists inside the glass cylinder, since there is no air in the tubes to carry away the heat. The absorber plate in each tube is coated with a selective coating. The Sun's radiation is absorbed by the selective coating and it is

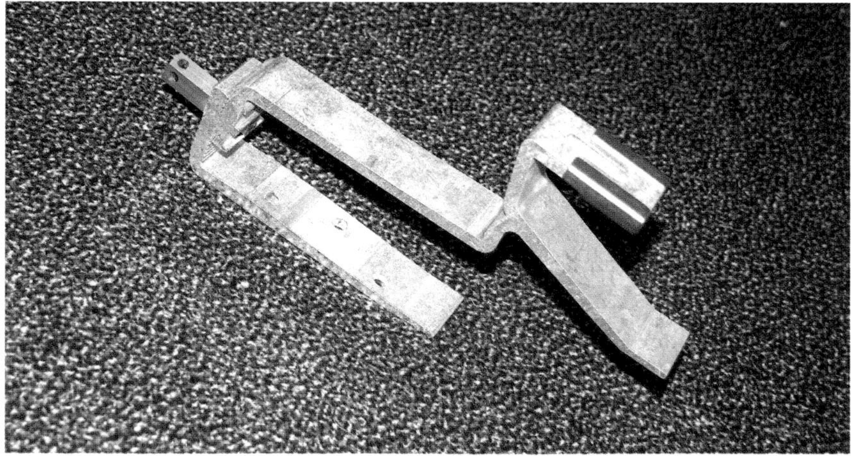

Figure 14: The aluminium bracket affixes to a wood member in the roof structure (lathe or rafter) and forms a solid fixing base for the panel support frame.

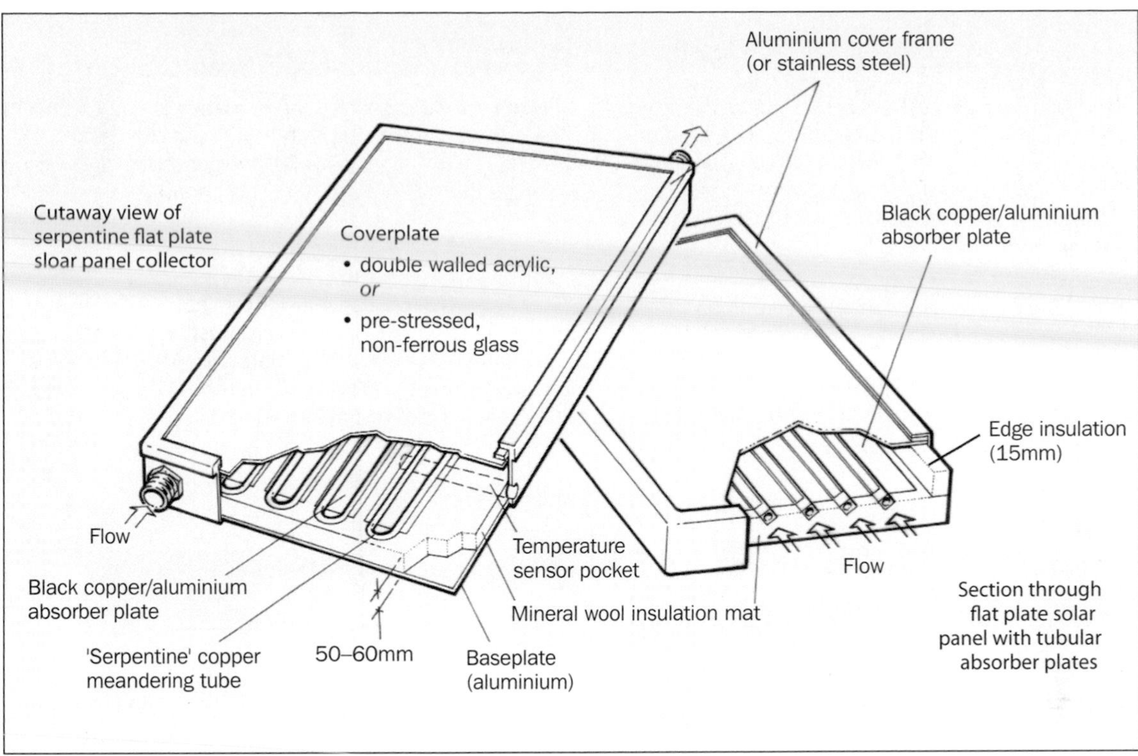

Figure 15: Flat plate solar collector.

prevented from re-radiating by the silvered coating on the inside surface of the tube. The tube acts like a one-way mirror that has been optimized to trap and absorb the infrared radiation. The glass tube also protects the absorber from moisture, condensation and any pollutants present in the air.

The absorber plate is soldered or braised to a heat pipe that holds a special fluid. Heat pipes take advantage of the thermal properties of some fluids, which have the capacity to hold and transfer large amounts of heat when they boil. The fluid is held inside a hollow tube (the heat pipe) at a pressure chosen so that it will boil at the hot end and condense at the cold end. The advantage is that the thermal conductivity of a heat pipe is many times greater than an equivalent block of solid metal. Heat pipes can transfer large amounts of heat for small temperature rises. The working fluid inside the heat pipe, in turn, transfers its heat at the cold end to a header pipe or insulated manifold at the top of the array of individual collector tubes. The primary circuit flows through the manifold and this

Figure 16: Thermomax 20 evacuated tube solar panel installed on a tiled roof. (Author)

Evacuated Tube Collectors

The accompanying diagram shows the principal components of an evacuated tube. Glass tubes around 2mm thick are used, sealed at both ends to preserve the vacuum inside. If the vacuum is lost inside the tube and air enters, the absorbers in some tube designs change colour from black to pink or blue. This colour change can be seen from the ground. Other tubes fog or mist up and failed tubes, which do not carry frost, sometimes stand out from their neighbours in frosty conditions.

The absorber consists of a narrow strip of copper (50–100mm wide by about 2m long) with a selective black-coloured coating bonded to a copper tube. The tube may be a heat pipe or it might simply be a copper tube with the heat transfer fluid flowing through it.

The heat pipe is generally two concentric copper pipes, typically a 15mm outer pipe with a 10mm diameter inner pipe.

The heat pipe fluid flows through the space between the concentric pipes, picking up heat from the absorber strip and returning through the inner pipe. Each heat pipe transfers its heat to the primary circuit flowing in the manifold.

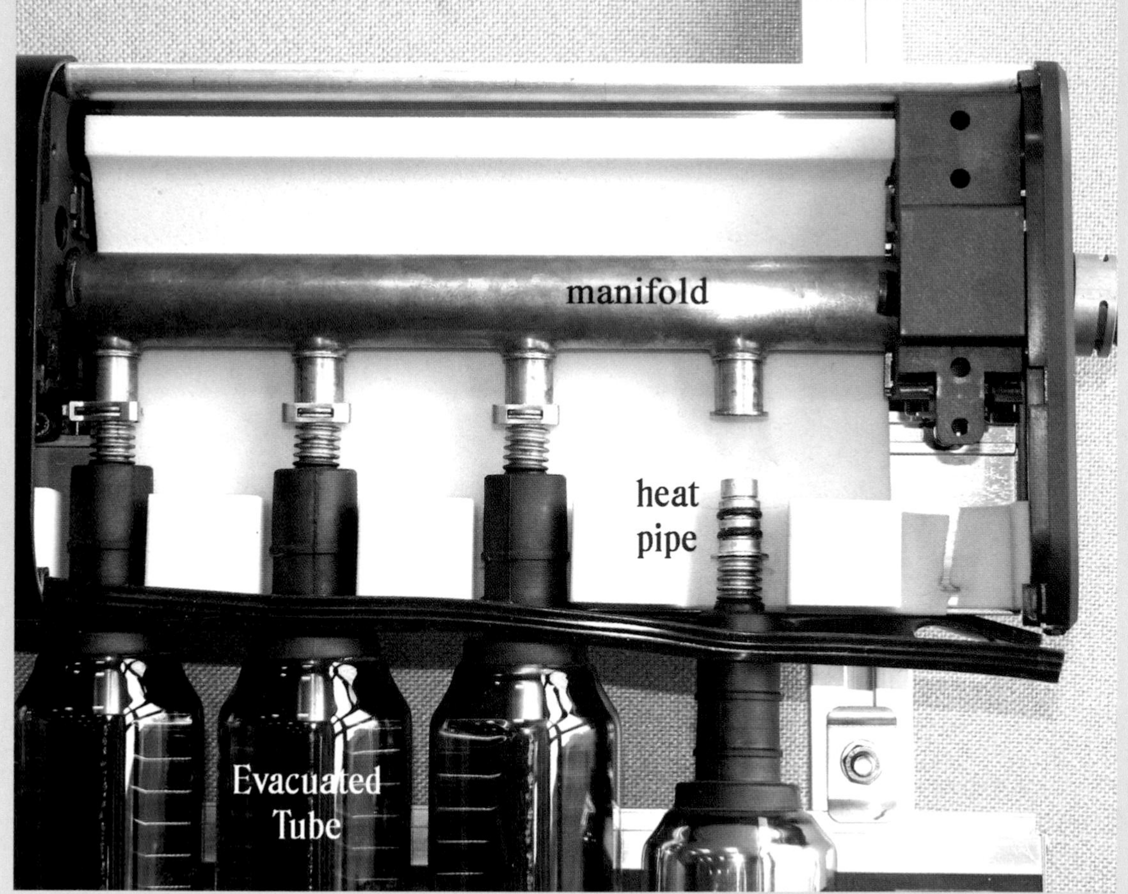

Figure 17: Detail of a Thermomax evacuated tube panel showing tubes and the open manifold box. The heat pipe (indicated) locates in the manifold. The tube seals and anti-shock coupling to the manifold are also shown.

flow carries the heat away from the solar panel to the hot water storage tank.

The absorber assembly is sealed within the evacuated tube, which is, in turn, coupled to the manifold. Each evacuated tube has its own connection to the manifold. In some evacuated tube designs individual tubes may be removed (for replacement in the event of failure) from the manifold without losing the heat transfer fluid from the primary circuit. To avoid thermal stress at the glass–metal junction, thermal shock absorbers are built into the glass–metal seals to help minimize the effect of rapid temperature variations. Evacuated tube assemblies usually comprise twenty or thirty tubes as standard units, although customized manifolds are available. The manifold is insulated and sealed in a weather-resistant cover. The bottom of each tube is supported by the framework to prevent flexure at the top joint.

If exact south orientation is not possible, since the panel can only be mounted on a roof facing south-east or south-west, the evacuated tubes may be turned in their mountings so that their absorber plates are facing south, giving an improved performance.

Absorbers

The absorber is a common element to both the flat plate and the evacuated tube collector and performs exactly the same role in both. It is a flat surface onto which the solar radiation falls. Absorbers are similar to central heating radiators in reverse; instead of emitting heat, they absorb it. Central heating radiators painted black have been used effectively as absorbers in simple, DIY SWH systems.

The job of an absorber is to capture as much of the incident radiation as possible. Dark-coloured matt surfaces are the most effective way to do this. Absorber plate surfaces are coated with selective or non-selective coatings. Non-selective surfaces, such as those painted matt black or dark navy blue (or they may be black plastic), absorb around 95 per cent of the incident radiation and they are also good emitters of radiation (losing around 90 per cent).

Selective surfaces are also good absorbers of radiation (absorbing 90–95 per cent), but these surfaces are poor emitters of thermal radiation. This has the advantage that the efficiency of heat transfer remains high as the temperature of the absorber increases, which is not the case for non- oxidized stainless steel or nickel in aluminium oxide, and can increase the energy captured by a collector by as much as 20 to 25 per cent over a year.

The absorber must be resistant to attack from the heat transfer fluid and able to withstand its own maximum stagnation temperature – the temperature that the absorber could reach on a sunny day with no fluid flowing in the collector. In glazed and insulated collectors, absorbers are made from copper, aluminium or stainless steel. Copper is more expensive, but it is a better heat conductor and is less prone to corrosion than aluminium or steel.

To avoid hotspots, the absorber plate should make full contact with the heat transfer fluid (it needs to be fully wetted), so that heat can be carried away from the rear of the plate in an even fashion. A stainless steel absorber plate, although heavier than other materials, provides an excellent solution. The tube-in-sheet (or tube-in-fin) design uses high conductivity copper or aluminium and provides excellent absorption characteristics. A single tube-in-fin design is used exclusively in evacuated tube designs. It is important that the collecting surface and the tube are well bonded so that a good transfer of heat from the surface to the fluid inside the tube takes place.

The heat transfer fluid is introduced to the collector at the bottom of the panel and, since hot fluid rises, it flows through an array of parallel channels or risers to the top. In some panel designs it can flow through a twisted path of pipework known as a serpentine. If the parallel channels are used, they will connect to common pipes, called headers, at the bottom and the top of the collector. The headers have a larger diameter (and hence a larger cross-sectional area) than the channels, ensuring there is a good flow distribution in each riser. Rapidly changing sunshine conditions require a fast response from absorbers to carry the heat away quickly, and this is best achieved with a small amount of fluid. A good absorber will have a fluid capacity of around 2 litres/m^2.

SWH STORAGE TANKS (HOT WATER CYLINDERS)

In most SWH installations to an existing building (or retrofit installation), the existing hot water storage

Table 8: Recommended hot water storage tank volumes (flat plate collector).		
Collector Area (m^2)	Hot Water Storage Tank Volume – Single Coil (Litres)	Hot Water Storage Tank Volume – Dual Coil (Litres)
3	135	150
4	220	190
6	250	280

cylinder should be replaced with a specially designed solar-insulated, twin-coil cylinder of larger volume. The twin coil means that the tank may also be heated from the primary heating. As described previously, an alternative to the coil is to use a separate pre-heat tank, although this is less efficient (by 2–3 per cent).

Figure 18: Installation with a 50mm insulation jacket, 180-litre hot water cylinder, circulating pump and Thermomax pump controller. Note also the cold water connection, using the cold water main, which allows heat transfer loop pressure to be set at around 2Bar.

Whichever design is selected, extra space will be required and this may be the determining factor. It is also prudent to remember that there is additional weight in the extra water, and if a larger cylinder is used then the floor below it must be capable of supporting the additional weight. In tanks with two or more coils it is important to remember that the lowest coil should always be used for the SWH. This is to enable the solar heat exchange coil to transfer heat to a large volume of water over a long period, during, for example, a sunny summer day.

SWH storage tanks should be insulated with at least 50mm of high-density polystyrene or fibre glass insulation. The volume of the storage tank fitted with a SWH system will be larger to provide greater quantities of stored hot water to allow for periods of bad weather. In most retrofit installations the normal tank, holding about 100 litres of water, is replaced with a larger tank holding 160 or 220 litres and with at least two internal heating coils. The hot water tank will usually have an immersion-type electrical heater fitted (say 3kW).

Some installers use a hot tube coil, which screws into the existing immersion heater flange. This means that the solar heating coil is located at the top of the existing cylinder and cannot heat the tank from the bottom. It is a cheap option that can be considered where tank replacement is difficult.

Under normal usage, the tank temperature reached as a result of solar heating alone will be lower than it would be when the storage tank is heated from an immersion heater or from a central heating system. As a rule of thumb around 50 litres of water storage is required for every square metre of collector area installed. For an average family this is around 200 litres. Increasing the storage volume will increase the system efficiency but will give a lower storage temperature for any given solar conditions. Storage capacity should never fall below 35 litres per square metre of collector area since dangerously high temperatures could result. Correspondingly, an average domestic hot water consumption for a person is around 40–45 litres per day, at an average temperature of 55°C. For a family of four that will be 4 x 45 litres = 180 litres.

A recent innovation is the Willis Solasyphon heater. The heater is connected to the solar collector in the

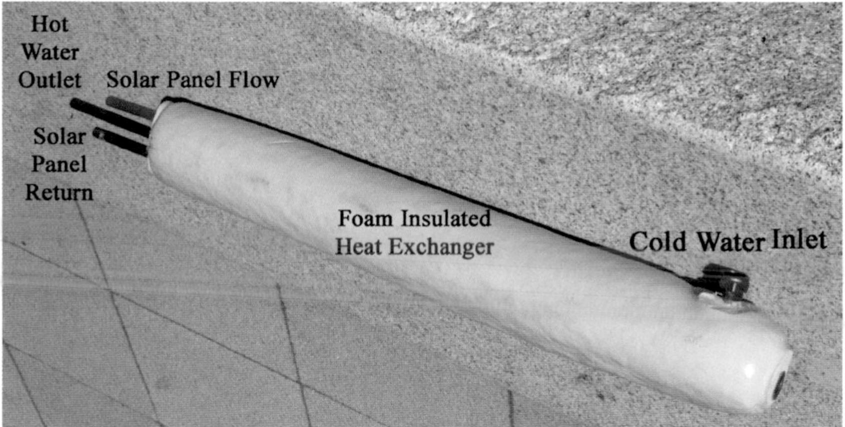

Figure 19: The externally mounted Willis Solasyphon brings the advantages of almost instant hot water, easy installation and cost savings, since the existing hot water cylinder is retained.

normal way, but it uses a heat exchanger to transfer solar-heated water rapidly to the hot water draw-off pipework, rather than to a coil in the hot water storage tank. It consists of a highly insulated outer shell, with an internal parallel tube heat exchanger. The outer tube is connected to the existing copper cylinder at the lowest possible point to provide the cold feed to the Solasyphon. The syphon outlet pipe is connected to the hot water draw-off pipework from the existing cylinder. Within the outer shell is a double-pass parallel heat exchanger. The other two protruding tubes shown in Figure 19 are the flow and return pipes from the solar collector.

The Willis Solasyphon offers the advantage that even in marginal or intermittent sunlight conditions the heater will provide useable hot water rapidly, which is not possible with existing systems. A conventional solar water heat transfer coil will take some time to raise the temperature of all the water in the storage tank. Another advantage of the Willis is that the existing hot water cylinder can be retained. The Solasyphon connection is straightforward and so the disruption and full cost of a standard system can be avoided. Savings in labour and materials mean that about a 30 per cent reduction in installation costs can be achieved. If hot water is not drawn off from the Willis Solasyphon in intermittent slugs, then natural thermosyphonic circulation will eventually raise the water temperature in the hot water storage tank in the normal way. Early successful trials have indicated that the product has the potential to make a significant impact on the market.

FITTING A SWH SYSTEM TO VENTED AND UNVENTED SYSTEMS

Normally most building hot water raising and distribution systems are vented, which means that the build-up of pressure in the system, as water temperature rises and the water expands, can be vented safely to the atmosphere. These systems generally have a cold water storage tank and are open to the atmosphere. There are, however, an increasing number of unvented (or pressurized) systems, where the system is effectively sealed and special attention must be given to how the temperature and pressure in the system are safely controlled. In an unvented system this necessitates a vessel, known as an expansion vessel, that can accommodate the expansion. In these systems, temperature relief valves and air vents must be fitted to prevent dangerously high pressure build-up. The release of scalding fluids and vapour from these vents and relief valves must be safely routed outdoors at ground level.

In a SWH system the primary heat transfer circuit is pressurized, usually from the town's water main, to around 1.5 to 2 Bar to prevent the fluid from boiling and to increase the system efficiency – remember water boils at 100°C (at 1 Bar) – and water temperatures can reach much higher than that in the primary circuit. The circuit requires an expansion vessel and pressure relief valves to be fitted. The expansion vessel is incorporated to prevent damage to the system when the volume increases as the temperature of the heat transfer fluid (the water/glycol mixture) rises. In its normal form, glycol is poisonous, so SWH systems must use edible glycol, which is safe should contamination of potable water occur. The expansion vessel has an internal high temperature-tolerant, rubber diaphragm, which separates the vessel into two chambers. One chamber is exposed to the heat transfer fluid in the primary circuit and the other is filled with air. As heat increases the fluid volume, the diaphragm expands and is displaced against the air-filled chamber. For most domestic systems a 4-litre expansion vessel is adequate. Expansion vessels should only be fitted to the return pipe run of the primary circuit to avoid overheating of the vessel.

COMBINING SOLAR WATER PRE-HEATING WITH A COMBI-BOILER

Unfortunately most combi-boilers that have already been installed cannot accept a feed from a solar water pre-heater storage tank. Feed water at a temperature greater than 20°C will result in boiler shutdown due to overheating. These boilers do not automatically reset, so an engineer's visit is required. However, there are some boilers where it is possible to connect the output from the combi-boiler to the solar water pre-heat tank. Care must always be taken when connecting SWH to combi-boiler systems.

OTHER SYSTEM COMPONENTS

Controller

Indirect SWH systems are fitted with differential temperature controllers. These are necessary, as explained previously, to use the pump to circulate the hot water in the primary circuit from the collector to the storage tank. A flow meter is used to measure the flow in this circuit and should cover the range of 4 to 15 litres per minute. In most instances a higher temperature of around 6°C to 8°C in the collector is required before the pump switches on and the controller will not switch off the pump until the temperature in the collector drops to a preset differential, usually 3°C. In most instances there is also a time delay in the pump circuit, which means that it will not switch off again for at least two minutes after it has switched on. This is to prevent 'pump hunting', which can damage the pump and waste electricity. In the vast majority of systems the factory settings for the controller will work well and will not need to be changed.

Some electronic controllers can provide a range of indications, such as the energy produced by the system over the course of a day in kilowatt hours. More sophisticated controllers might handle a range of duties including data-logging, safety monitoring, providing thermostatic and time-clock control of

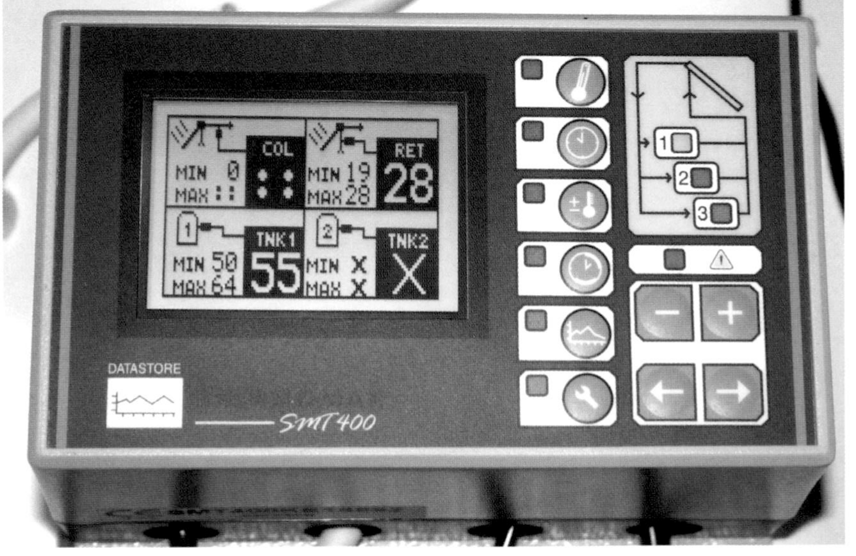

Figure 20: Thermomax system SMT 400 pump controller and temperature display unit.

Table 9: Preferred materials to be used in the primary circuit between the collector and the hot water tank.		
Pipework	**Conditions of Use**	**Preferred Fittings**
Black steel pipes	Can be used in all situations	Threaded
Copper pipes	Can be used in all situations	Compression
PEX pipes and ALU plastic pipes	Never to be used in primary circuit	
Stainless steel pipes	Can be used in all situations	Compression

auxiliary heat, monitoring hot water circulation loops or hot water storage tank temperatures, and displaying or transferring error messages or alarms. Sometimes they can also communicate with remote displays or computers.

Thermal Mixing Valve

SWH systems can produce water temperatures well above those normally experienced in domestic hot water systems, so it is good practice, if not essential to prevent scalding, to include a thermal mixing valve (or anti-scald valve) for the hot water take-off.

Pipework and Fittings

Systems should be installed to the highest plumbing and electrical standards and SWH systems do require particular attention with respect to pipework and fittings. Table 9 indicates the preferred materials to be used in the primary circuit between the collector and the hot water tank. Note that all gaskets and sealants must be temperature resistant, and galvanized pipes must not be used in loops containing antifreeze.

Plastic pipes should not be used in the primary circuit pipework because they may have to hold the water/glycol mixture, under high pressure, at temperatures of around 150°C. Most plastic piping can stand temperatures up to 120°C for short periods only and should not be used. Soldered joints (soft plumbing solder as opposed to high temperature welding or braising solder) must not be used in the primary circuit. Compression fittings (bends, elbows, straight couplers, etc.) and compression valve connectors should be used and joints sealed with plumber's flax (rather than PTFE tape) to prevent leakage, particularly at the hot tank cylinder where the pipework from the solar collector joins the tank.

Valves and fittings with plastic components or inserts should never be used in the primary heat transfer circuit. Experience has shown that one of the most common faults in systems is leaks (sometimes slow or recurrent leaks) from joints. This leads to loss of pressure in the primary circuit. Even experienced installers are not immune to these problems. High-quality compression fittings are recommended at all times and some experienced solar installers will also use high temperature jointing paste or plumber's flax to ensure a good seal at joints. It is essential that all pipework, joints, gaps between components and where the pipes pass through ceilings in the primary circuit are insulated at least to a thickness of 20mm of mineral wool or expanded polystyrene.

INSTALLING A SWH SYSTEM

Detailed instructions for installing a SWH system are outside the scope of this book since there are a range of systems on the market and each supplier's equipment will have a unique set of installation requirements. The following is suggested as a list of some key points to be considered when planning to install a system. In existing properties, the position of the hot water storage tank is probably fixed and there will be no opportunity to change the orientation and slope of the roof.

- Pick a location on a sloping roof that faces south. Most roofs slope at 30–45°, which is fine; orientations 30° east of south and 40° west of south are perfectly acceptable. If the roof slope is less than 20°, an additional collector area, perhaps 10 per cent extra, can be used to offset the loss due to tilt. Orientation away from south can be offset in the same way, but panels must never be installed on an east-, west- or north-facing roof.

Figure 21: Some manufacturers supply special roof integration and weatherproofing kits for building the solar panels into the roof during construction. This incurs an extra cost but saves on roofing materials.

- Having chosen a position for the collector, the pipe runs to the hot water storage tank should be kept as short as possible and well insulated.
- Ensure that the selected area on the roof does not fall under shadow from trees, buildings, chimneys, aerials, etc. Anticipate any potential shadow fall position at all times of day and over the four seasons.
- Ensure there is sufficient space and no obstructions to allow the necessary connections and the pipe runs through the roof.
- Ensure that the roof structure is sufficiently strong to support the additional weight of the solar panel, mounting frame and the pipe runs. For a flat plate solar panel the weight is usually distributed over several rafters. Strengthen the roof if there is any doubt. Wind lift should not be ignored, although this is probably only important in very exposed locations that receive strong gusts. Care is needed when handling flat plate collectors during installation because they have significant sail when caught in the wind and this can result in a strong unbalancing force.
- As noted above, panels can be mounted on flat roofs, affixed to gable walls or at ground level on angled frames (on concrete bases).
- Flat plate collectors have the cold water entering at a bottom corner and the hot water returns to the hot water storage tank from the top diametrically opposed corner. To avoid air locks in tube collectors, the panel can be tilted up slightly, highest at the corner where the hot water leaves. However, flat plate collectors with serpentine pipe flows should always be mounted without a tilt.
- Always use an accredited installer. When installation of a SWH system is due to take place, ensure that priority is given to the health and safety aspects of access to the roof and appropriate handling of the panels. Flat plate panels, while not unduly heavy, are more difficult to lift to the roof than evacuated tube collectors and even in light winds they can require additional care, owing to their greater area. For these reasons secure scaffolding and a stable work platform must always be in place. Evacuated tubes are made of glass under vacuum and so they should be carefully handled to avoid the risk of implosion. Appropriate protective clothing and goggles should be worn during installation and care must be taken to avoid damaging the tube-to-manifold seals during assembly.
- Once the panels are in place, the roof should be weatherproofed with flashing kits and mastic as appropriate. Special attention to sealing should be made where the pipework goes through the tiles or slates to the roof space. Roof integrated panels are normally only considered during new build construction and special flashing kits are available for most makes of panel.

- Water in the primary circuit can reach very high temperatures and even during installation the panel may produce scalding water.
- It is worth restating that it is essential to always use an accredited installer to do this work. There are excellent training courses available and lists of approved installers are detailed in the websites given on pages 25–27.

SWH EFFICIENCY AND SYSTEM ANNUAL OUTPUT

A typical flat plate installation in the UK might have a panel of three to four square metres, or two square metres for evacuated tubes, with a storage tank of 150 to 200 litres; the optimum size, however, will depend on actual hot water demand. Software is available that can simulate the system performance throughout the year and determine the most appropriate system sizing, depending on the site conditions.

The various collector designs and absorber surfaces considered previously give different efficiencies. In the range of temperatures over which the normal domestic hot water systems operate, there is little difference in performance between selective surface flat plate collectors and evacuated tube collectors. The options for the various panels and their collector surface coatings may be summarized as:

- Selective surface coating on evacuated tube panel.
- Selective surface coating on a flat plate solar panel.
- Non-selective surface coating on a flat plate solar panel.
- Unglazed non-selective surface coating on a flat plate solar panel (such as would be used to heat a swimming pool).

Figure 23 (overleaf) gives a first approximation of the output from these solar collectors. As an example, if the required water temperature is 60°C, then the curve indicates that an efficiency of around 70 per cent is obtainable from either a selective surface flat plate or from an evacuated tube collector. Assuming the incident radiation on a sunny day is around 1,000W/m^2 (see above), then each square metre of solar collector will be generating around 700W. For an average family home with 4 square metres of collector area, that will give around 2.8kW capture. To allow for a truer representation, taking into account factors such as atmospheric pollution, the ground level radiation is more likely to be around 850W/m^2, so the actual system output is going to be closer to 2.38kW.

An average home requires 3,000 to 4,000kWh per year of hot water. A SWH installation of 3–4m^2 of flat panel area or an evacuated tube system of twenty to thirty tubes will provide 1,800–2,000kWh of heat energy in a year and this is roughly 50 per cent of the energy needed to provide the annual hot water demand. This is known as the solar fraction. At mid latitudes, during June and July, a SWH system can provide almost all of an average family's hot water requirement. This reduces to 95 per cent in May, then to around 75–80 per cent in April, August and September. In the months of March and October the figure reduces to around 55–60 per cent and in the winter months of November, December, January and February less than 10–20 per cent of the requirement can be delivered. Even on grey, cloudy days, collectors can raise the hot water cylinder temperature by a few degrees. The effectiveness of a solar panel will depend heavily on the hot water draw-off frequency. Some homes with young families and a

Figure 22: Solar panel mounted on a flat roof showing the supporting frame attached to concrete ballast. This form of mounting allows full flexibility in orientation. (Photograph copyright Mark Compston)

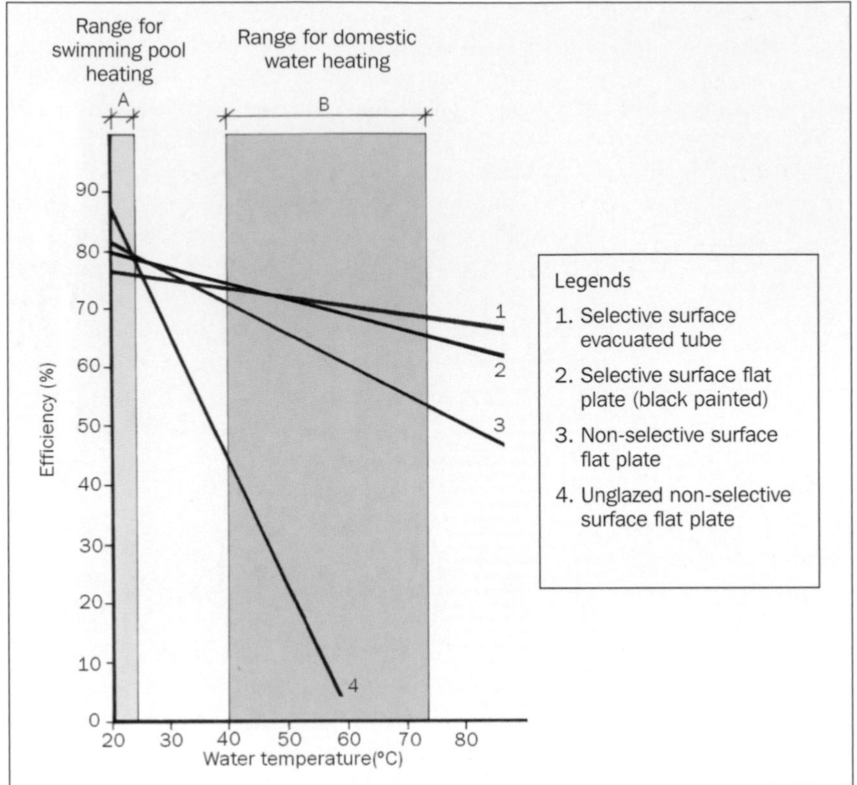

Figure 23: Operating efficiency range of solar panels with different absorber surface coatings. (Reproduced courtesy of the Renewable Energy Installer Academy [Action Renewables & SEI])

parent at home all day will have an almost continuous, high demand for hot water. Other homes with fewer occupants, who may be at work during the day, will require a lot less hot water.

SWH SYSTEM ECONOMICS AND CARBON SAVING

SWH systems can be considered as either low carbon or zero carbon systems, depending on how the pump is operated and controlled. If the heat transfer fluid is circulated by an electrically powered pump fed from mains electricity, it will be a low carbon SWH system. The use of electricity typically reduces the carbon savings of a system by 5 to 10 per cent. The cost of running the pump from mains electricity in a SWH system is around 5 per cent of the annual savings, but this will vary depending on a range of factors. Some systems have alleviated this problem and approach a zero carbon solution through the use of a pump driven by a small photovoltaic (PV) panel (typically 5–20W), which is mounted integrally or close to the water heating panels. This panel should face in the same direction as the main solar

Table 10: Daily solar energy reaching each square metre of the Earth's surface (approximate figures at latitudes around 50° north or south).

	Jan	Feb	Mar	Apr	May	Jun	Jul	Aug	Sep	Oct	Nov	Dec
(kWh/m^2)/day	0.6	1.2	2.0	3.3	4.2	5.0	4.5	3.6	2.8	1.7	0.8	0.5

For comparison, solar collection area: 20-tube panel = $2.25m^2$ / 30-tube panel = $3.38 m^2$.
(NB: these figures are only indicative; any one month can receive an unusually large or small amount of radiation).

Figure 24: These Aton Imagination solar panels have a small PV module (on the right-hand side), which is used to power the SWH pump. (Photograph copyright Mark Compston, 2007 www.imaginationsolar.com)

heating panel and be used to drive a small, low-flow diaphragm type pump.

At current prices (2008) a typical fully installed SWH system would cost around £3,000 to £4,500, depending on such factors as the equipment make and size to be installed, the installation cost (including scaffolding) and the panel location. Once installed the operating costs are small, since the heat energy is free, but, as noted, some energy is required to drive the pump. Payback is used as an indication of the economic success of a SWH. Although every situation is different, most systems will offer economic paybacks of around fifteen to twenty years, which will be reduced if a capital grant is available. Allowance must be made for potential increases in fossil fuel prices and a full consideration of the impact on the main heating system could be included.

Grants are a key factor when the installation of a SWH system is to be considered. These have been available at around 10 to 30 per cent of the total installed cost, although it is a widely held belief that SWH is now commercial and no longer requires government grant intervention. Grants have not been available for DIY installations. Government grant programmes are continually under review and in the past they have been changed or withdrawn at short notice. (For details of the relevant grant websites, *see* pages 25–27.)

With SWH, the central heating boiler's lifespan will increase because it has to do less work to get the water up to temperature and components will last longer without failing. Properly installed and maintained SWH systems should last in excess of thirty years. A primary consideration must be the reduction in carbon dioxide produced by offsetting fossil fuel consumption. This is estimated at 0.3 to 0.75 tonnes per year, giving around 20 tonnes over the lifetime of the system. These factors are not included in the payback calculation and may be critical, particularly since it is difficult to put a value against the carbon saving.

PLANNING AND BUILDING CONTROL REQUIREMENTS FOR SOLAR PANELS

In most cases planning permission will not be needed for roof- or wall-mounted solar panels if they do not project beyond the existing surface plane of the roof by more than 150mm and do not exceed the height of the highest part of the existing roof.

Different conditions apply if the panels are to be erected elsewhere within the boundary of a property, for example on the ground or on a new, dedicated building. If the panels are to be fitted to a listed building, or if the building is in a conservation area, then planning permission may be required and a fee will be charged, but local conditions vary and the local planning authority should be contacted before installation is commenced. Government planning guidance for the planners exists in the UK and in Ireland, and it is there to facilitate and encourage the greater use of renewable technologies, such as solar panels.

SWH panels generally require a Building Control Certificate, issued after an inspection by the local Building Control Officer. The fee for this inspection can vary from region to region. Building Control approval is required and should ensure that the installation is safe and will not cause structural problems. It will also form an integral part of the energy rating of the building and the Building Control Certificate issued will be included in the Home Information Pack (HIP) for the building.

Table 11: The golden rules of SWH	
Top Tips for SWH	**Things to Avoid With SWH**
Make sure the SWH panels are orientated as close as possible to south and tilted at an angle of approximately ten degrees less than your latitude angle.	Do not install SWH panels on roofs that face directly west, east or north. Panels can be mounted on the ground or on flat roofs using suitable orientated and tilted mounting frames.
Use approved equipment and accredited installers. Seek the views of more than one installer and customer.	Do not expect the same quantities of hot water in winter, or during periods of prolonged cloud cover, as are produced on bright summer days.
If you occupy a listed building or live in a conservation area ensure you have early contact with the planning authority about the proposed SWH installation.	Do not place SWH panels where they would be shaded by trees, buildings or other structures – allow for shadow-fall throughout the day and across the seasons.
SWH panels normally require a Building Control Certificate and inspection, so contact the local Building Control office before work is commenced.	Do not consider SWH panels as a means of providing space heating unless there is a well-proven, year-round solar resource. SWH can pre-heat feed water in a boiler heating system.

SWH GLOSSARY

Combi-boiler system A gas-fired boiler system that provides hot water and radiator heating on demand but does not include hot water storage.

Drain down Means that the heat transfer fluid (water, brine or a water/glycol mixture) in the primary circuit, which includes the pipes in the collector, will flow back into a tank without being replaced, leaving the collector and the primary circuit pipework with no heat transfer fluid in it.

Fossil fuels Coal, heavy fuel oil, gas, peat, lignite, distillate oil (central heating oil).

Hard water areas Regions of the country that draw their water from sources rich in minerals. Hard water usually contains ions of calcium (Ca^{2+}), magnesium (Mg^{2+}), and possibly other dissolved substances such as bicarbonates and sulphates.

Insolation The amount of solar radiation or light received from the Sun.

Latitude angle The angle between the plane of the Earth's equator and the position of the site on the globe. The latitude angle is the value of the latitude in degrees.

Natural circulation A process of convection heat transfer that occurs when a heated gas or fluid rises and cooler air flows in to take its place. The movement of the fluid or gas is a circulation. This occurs naturally because the heated fluid has a lower density than the cooler fluid. The hot fluid rises, cools and sinks only to be heated and rise again in a circulation process.

Pre-heater tank A storage tank that can be used to collect hot water in addition to the main hot water storage cylinder, and which can be used to store water, sometimes at a higher temperature than the main tank.

Space heating Heating provided for houses, commerce and industry usually derived from a boiler with radiators.

Thermal mass The amount of heat that any given material may hold.

USEFUL SWH CONTACTS AND WEBSITES

SWH Information and Books

Low Impact Living initiative (LILI)
www.lowimpact.org/acatalog/books_solar_hot_water.html
Good range of solar energy publications.

IEA Solar Heating & Cooling Programme
www.iea-shc.org/
Reports on a range of solar projects carried under the auspices of the International Energy Agency.

Useful Website Links

Arcon
www.arcon.dk
Danish company specializing in large-scale active solar heating systems.

Energy Efficiency Best Practice in Housing
www.energysavingtrust.org.uk/housingbuildings/professionals/
Energy Saving Trust site providing a large number of case studies of low energy and passive solar housing projects.

Energy Research Group, University College Dublin
http://erg.ucd.ie
Details research, particularly on low-energy and passive solar building design.

European Large-Scale Solar Heating Network
www.enerma.cit.chalmers.se/cshp/
Based in Chalmers University of Technology in Sweden, the network concentrates on active solar heating for large buildings and district heating.

European Solar Thermal Industry Federation (ESTIF)
www.estif.org
Trade organization promoting active solar heating throughout Europe.

International Solar Energy Society (ISES)
www.ises.org
International Headquarters, Villa Tannheim, Wiesentalstrasse 50, 79115 Freiburg, Germany
Information on ISES activities, conferences and publications, including *Renewable Energy Focus* magazine.

Mysolar – IEA Solar Energy Site
www.mysolar.com
Independent information on solar energy and solar systems: products, prices, sales outlets, DIY design and selection, solar electricity systems and solar water heaters worldwide.

National Energy Foundation
www.nef.org.uk
Good source of SWH information.

Solar Trade Association Ltd
www.greenenergy.org.uk/sta/
Davy Avenue, Knowlhill, Milton Keynes
MK5 8NG
Trade organization promoting solar heating in the UK.

Thermomax Ltd
www.thermomax-group.com
7 Balloo Crescent, Balloo Industrial Estate, Bangor, Co. Down BT19 7UP, Northern Ireland
Company specializing in the manufacture of evacuated tube solar water heaters.

UK Health and Safety Executive
www.hse.gov.uk

UK Solar Energy Society
www.uk-ises.org
UK section of the International Solar Energy Society.

CHAPTER 3

Heat Pumps

We need to stop living in poorly insulated, high heat-loss homes using radiators containing water at 70°C and start constructing super-insulated buildings that are heated using water at 35°C, through under-floor or wall heating systems.

HEAT FROM THE GROUND

Heat pumps have been in use in Canada and North America for around seventy years, and in Europe for more than forty years, but in the United Kingdom and Ireland they are probably the least well known of the renewable technologies. In our modern lives we are surrounded by heat pumps in the form of fridges, air conditioning units and fresh-food vending machines. In this chapter we will look at how heat pumps provide space and water heating by drawing low grade heat energy from the environment, which is then raised to a more useful temperature through compression, using electricity. This low grade or low temperature heat can come from any ambient source, that is, the air, water in the form of a stream or lake, or the ground just below the surface. Around one quarter of the heat energy produced by the heat pump is provided by electricity, and the other three quarters is drawn from the environment.

Ground source heat pumps are sometimes referred to as geothermal heat pumps and this infers a link to the geothermal energy available in the rocks and waters heated from the Earth's interior magmas. The majority of the heat pumps under consideration in this chapter draw their heat from the Sun in the form of solar energy stored in the upper regions of the Earth's surface. For most ground source heat pump systems, the heat resides in the top one or two metres below the ground. Vertical borehole heat pumps are an option where space restrictions or ground conditions make horizontal systems unsuitable. Vertical borehole collectors are more expensive than horizontal collectors. They draw their heat from both the deeper, true geothermal resource (*see* Chapter 8) as well as from stored solar energy. The vertical collectors on these heat pumps go down at least 60 to 100 metres into the ground, and they generally provide higher efficiencies.

Although heat pumps use electricity, they are classified as renewable because they save on fossil fuels and the electricity they consume can come from a renewable source, such as a photovoltaic array or a wind turbine. Properly designed and installed heat pumps offer a very low carbon dioxide emission option for space heating. Less relevant in our climatic conditions, heat pumps can also provide cooling as well as heating and they can save energy costs of around 30 to 70 per cent in heating mode and 20 to 50 per cent in cooling mode, compared with conventional systems. Environmentally, they are benign, producing no on-site pollutants, and they have the lowest harmful gas emissions of all the heating or cooling technologies.

HEAT PUMP HISTORY

William Thomson, later Lord Kelvin, first developed the idea in 1852 and the concept became a practical reality in the 1940s when Robert C. Webber, an American inventor, noted that the outlet pipe of his deep freezer almost burned his hand. He immediately identified the opportunity for heating water with this waste heat. He developed his discovery further to include space heating by using the waste heat from the freezer to warm air, which was then circulated around his home using a full-scale heat pump.

Webber also noticed that the ground below the surface maintained a reasonably stable temperature throughout the year, and he suggested that it could act as a source of low grade heat. His early design involved a freon-filled copper tube running through the ground to pick up the heat. This formed a heat exchanger, which he used to heat air. His house was heated by circulating the resulting hot air through the rooms with a fan.

Heat pumps came to the fore during the Middle East oil crisis of the 1970s, when a sudden need for greater conservation and a heightened concern for security of energy supplies led people to consider alternative technologies. The concept was taken up by Dr James Bose, at Oklahoma State University (OSU), who redesigned the heat pump into a more useable device. OSU has since become a global centre of excellence for heat pump research and development. The market for heat pumps in the UK and Ireland is currently small but this is increasing, with the domestic and commercial markets dominating.

HOW A HEAT PUMP WORKS

To understand the workings of a heat pump, it is useful to remember that all our surroundings – the air, the ground, water in lakes and ponds and even a block of ice – contain heat. The purpose of a heat pump is to absorb heat where it is plentiful and to transport and release it in another location where it can be used for space and/or water heating. It can be described as pumping heat from one place to another rather than creating new heat, as is the case with a combustion boiler. Heat flows from a high temperature to a low temperature and an example is the heat flowing from the heated interior of a house to the outside surroundings through the walls and windows.

Heat pumps operate by circulating cold water (or a water-antifreeze mix, referred to as brine) through collector pipes immersed in a heat source. The temperature of the water in the pipes as it leaves the heat pump to the collector is lower than that of the surrounding heat source and so it warms up by taking heat from the source.

The heat absorbed by the external collector is transferred to a heat exchanger in the heat pump. The heat exchanger contains a fluid, the refrigerant, which absorbs the incoming low grade heat. In some systems, known as DX systems (see below), the refrigerant is circulated directly in the collector circuit. Refrigerant fluids boil from a liquid to a vapour when heated, and when cooled they condense back into a liquid again. This property makes them very efficient at absorbing heat and releasing it. Heat pump refrigerants have low boiling points, typically, -25°C to -45°C at normal temperature and pressure. A typical heat pump refrigerant is R407 C, which has a boiling temperature of -43.9°C.

Once it has absorbed the low grade heat from the collector, the refrigerant is then compressed, which increases the refrigerant temperature to around 30 to 45°C. In some heat pumps this temperature can be much higher, say 55°C. The heat is then extracted from the refrigerant by a second heat exchanger (the condenser) to be used for space or water heating. The building and the heat delivery system should be designed to allow this temperature to be as low as possible, but clearly the design must provide adequate heating to meet the demand. The compressor in the refrigeration system produces waste heat, and in the most efficient designs most of this can be recovered, further reducing running costs and saving on carbon dioxide emissions.

The refrigerant, now much cooler, passes through an expansion valve to cool it further, dropping the temperature to around -1 to -3°C, making it ready to accept more heat from the heat source. A pump to circulate the refrigerant and a metering device to control its flow completes the system. The various components are all interconnected by pipes and self-contained within an enclosure no bigger than a large refrigerator.

Heat Pump Definitions: Seasonal Performance Efficiency and Coefficient of Performance

The efficiency of a heat pump system will depend on getting the heat source temperature as high as possible, while keeping the space and water heating temperatures as low as practicable. The ground temperature remains stable all year, while the air temperature is highly variable on a seasonal and daily basis, chilling rapidly in winter when heating demand is at a maximum.

Seasonal Performance Efficiency (SPE) is defined as the ratio of the energy delivered by the heat pump to the total energy supplied to it over a year (this is the total system demand including circulation pumps). Heat pumps that supply wet space heating systems can operate at seasonal efficiencies of around 300 to 400 per cent for indirect systems and at 350 to 500 per cent for direct systems (described below). Air source heat pumps by comparison operate with seasonal performance efficiencies of around 250 per cent. However, it must be remembered that it is the heat source and heat sink temperatures that dominate the heat pump SPE, but the design and extent of the building insulation is crucial. Generally heat pumps will result in a 45 per cent reduction in CO_2 emissions against a fossil-fuelled heating system.

Another measure of the efficiency of a heat pump is the Coefficient of Performance (CoP). The higher the CoP the better the system is performing. A typical residential heat pump system might be operating with a CoP of 3.4 (an efficiency of 340 per cent), which means that for every one unit of electrical energy (kW) supplied to the system, 3.4 units of heat energy (kW) are put into the home as heat. This compares to an energy-efficient condensing gas boiler,which will have an efficiency of 96 per cent or 0.96 (that is, less than 100

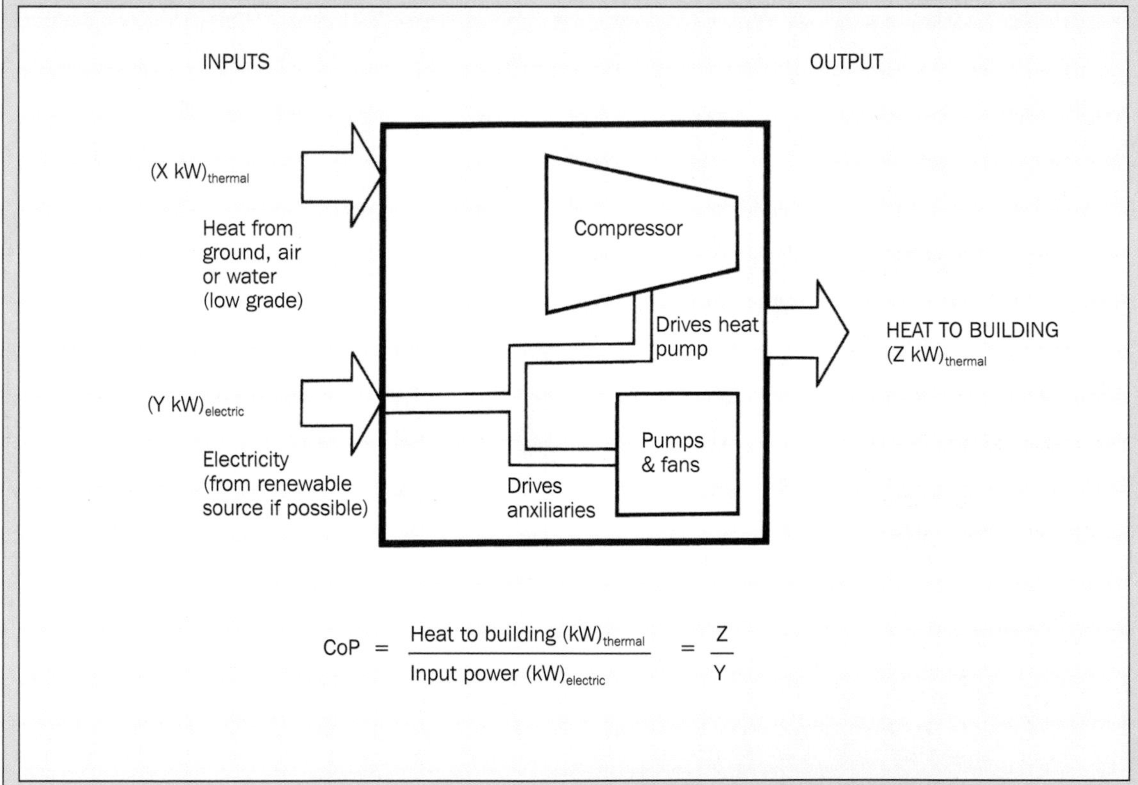

Figure 25: Coefficient of performance (CoP) is given by dividing the heat supplied to the building by the electrical energy input.

Heat Pump Definitions
(continued)

per cent). The CoP drops rapidly as the ground temperature falls: for example it could be 4.5 when the ground temperature is 15°C, falling to 3.4 for a temperature of 1°C.

An indication of efficiency when the heat pump is delivering cooling is the Energy Efficiency Ratio (EER). More efficient systems will have a higher EER.

Heat pumps that deliver a CoP of three to four are performing well. Heat pumps designed to deliver higher CoPs will generally be more expensive. An important principle is that a higher output temperature requirement will result in a lower CoP for all forms of heat pump.

The international performance standard for heat pumps is BS EN 14511, 2007 and this will set down the CoP and how it can be achieved for any given heat pump. However, its defined operational conditions are not always the ones that the system will experience, so care must be taken in interpreting this information. Heap pump standards are listed in Appendix II.

Heat Pump Vapour Cycle

The thermodynamic cycle for heat pumps may be understood as four separate processes that make up the Refrigeration or Vapour Compression cycle. The sequence is:

- **Evaporation:** Heat drawn from the heat source is used to boil and evaporate the refrigerant.
- **Compression:** The temperature of the refrigerant vapour is increased by compression, with the energy to drive the compressor drawn from electricity.
- **Condensation:** The heat is removed from the vapour in order to heat the building and the refrigerant cools and condenses back into a liquid.
- **Expansion:** The liquid refrigerant, which is now cool, is expanded to a low pressure by passing it through an expansion valve, which results in a further sharp fall in temperature. The refrigerant is now ready to pick up the low grade heat from the heat source (ground, water, air).

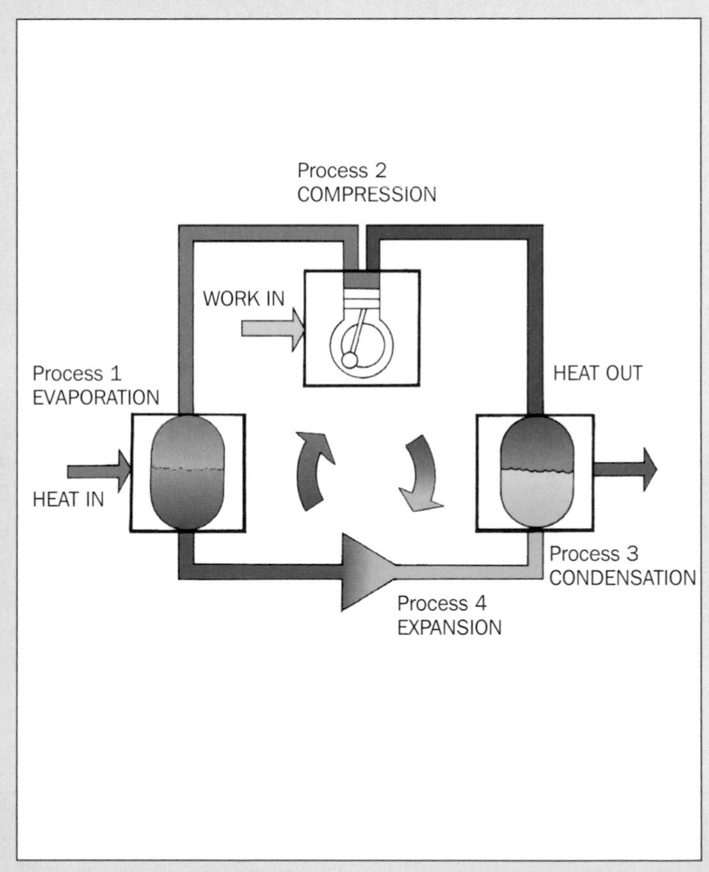

Figure 26: Heat pump vapour cycle – operating schematic.

HEAT PUMP COMPONENTS

Heat pumps operate on the basis of the refrigeration thermodynamic cycle, which calls for two heat exchangers: one that absorbs heat, the other that rejects it. The following is a more detailed technical description of the main components of a heat pump.

Compressor

The compressor and the thermostatic expansion valve maintain the correct operating temperature and pressure in the evaporator. The compressor draws the gas from the evaporator and compresses it to a high temperature and pressure. Non-return valves (NRVs or check valves) are installed after the compressor to prevent the gas flowing backwards, although these are sometimes integrated into the compressor. The most common type of compressor is the scroll compressor, although there are rolling piston compressors and simple piston compressors. Normally the electric motor and the compressor are mounted in the same vessel or casing, which is hermetically sealed. The leads required for the electrical connection pass through the seal.

Scroll compressors are the most common type of heat pump compressor because they have few moving parts and operate with minimal vibration and low noise. These compressors are relatively unaffected by small drops of liquid entering with the gas, although larger amounts will cause serious damage. Rolling piston compressors are most common in air-to-air heat pumps. They employ a rotating eccentric shaft with a radially moving vane to compress the gas. This provides a better CoP than simple piston compressors of a similar capacity. The electric motor windings of a rolling piston pump are often cooled by the refrigerant gas.

Simple piston compressors have been available for a long time. They rely on the reciprocating action of the pump working in sequence with valves to compress the gas. The latest developments in this type of compressor use special valves in the piston itself and this allows the CoP to be increased.

Expansion Valve

The expansion valve controls the flow of refrigerant between the condenser and the evaporator. It acts as a variable throttle valve, maintaining the pressure difference between the high- and low-pressure sides of the heat pump. As the refrigerant liquid passes through the expansion valve it immediately experiences a pressure drop and it turns into a gas; this is known as a phase change. The energy driving the expansion comes from the gas itself, so its temperature drops as the energy is taken out (as latent heat). The flow rate of the refrigerant through the evaporator is such that it is fully evaporated to prevent large amounts of liquid entering the compressor, since damage will be caused if the compressor attempts to compress fluid.

Expansion takes place through an orifice (a small hole or capillary) and control can be introduced by changing the size of the hole. With changing load conditions, this control can be accomplished automatically using a thermostatic expansion valve. The expansion valve contains a bulb filled with a material that responds to the changing flow conditions through the valve. The bulb acts with an electronic controller to provide the required design conditions accurately by opening and closing the orifice.

Evaporator

The evaporator is a heat transfer device, using the heat pick-up from the collector loop to evaporate the refrigerant. The constant draw of the compressor ensures that the temperature of the evaporator is always lower than the temperature of the heat source. In air source heat pumps the ambient air is exposed directly to the finned coils of the heat exchanger and this acts as the evaporator. These finned coils are copper tube arrays with aluminium fins pressed onto them, to increase the surface area and enhance the heat transfer. Flat plate heat exchangers are the most common in use today because they are economical, compact and easily insulated. The refrigerant leaving the evaporator is a superheated gas (its temperature is above the evaporation temperature) and this is drawn into the compressor.

Condenser

The condenser is used to transfer the heat available in the refrigerant into the heating system (which may be an under-floor heating system, radiators, wall-heating or ducted air) through a flat plate heat exchanger. Compact flat plate systems are commonly used and these require a liquid receiver since they have a limited storage capacity for refrigerant at times of low heating demand. The temperature of the refrigerant must be above the temperature of the space heating to enable the heat transfer to take place. The heat transferred is the total of the heat taken from the heat source in the evaporator and the heat equivalent of the electrical energy supplied to the compressor.

When the hot vapour enters the condenser it is desuperheated (or cooled) to the condensation temperature. The refrigerant then condenses back to a liquid and it is at this stage that the majority of the heat transfer to the space heating takes place.

There is a further stage in the cooling of the refrigerant to a liquid, called subcooling. This process lowers the temperature of the refrigerant to prevent the formation of vapour bubbles, which can block the refrigerant flow at the expansion valve.

The compressor, expansion valve, pump and refrigerant are sealed under pressure as part of the heat pump and these are not disturbed during the installation process. The refrigerant is normally sealed for life in the system unless a fault or leakage requires the system to be refilled.

The components act as described above in a continual process while the compressor is running and circulating the refrigerant. Heat pumps contain special switches that ensure the pressure conditions inside the pump are acceptable. A low-pressure switch or pressostat is designed to protect against leaks that

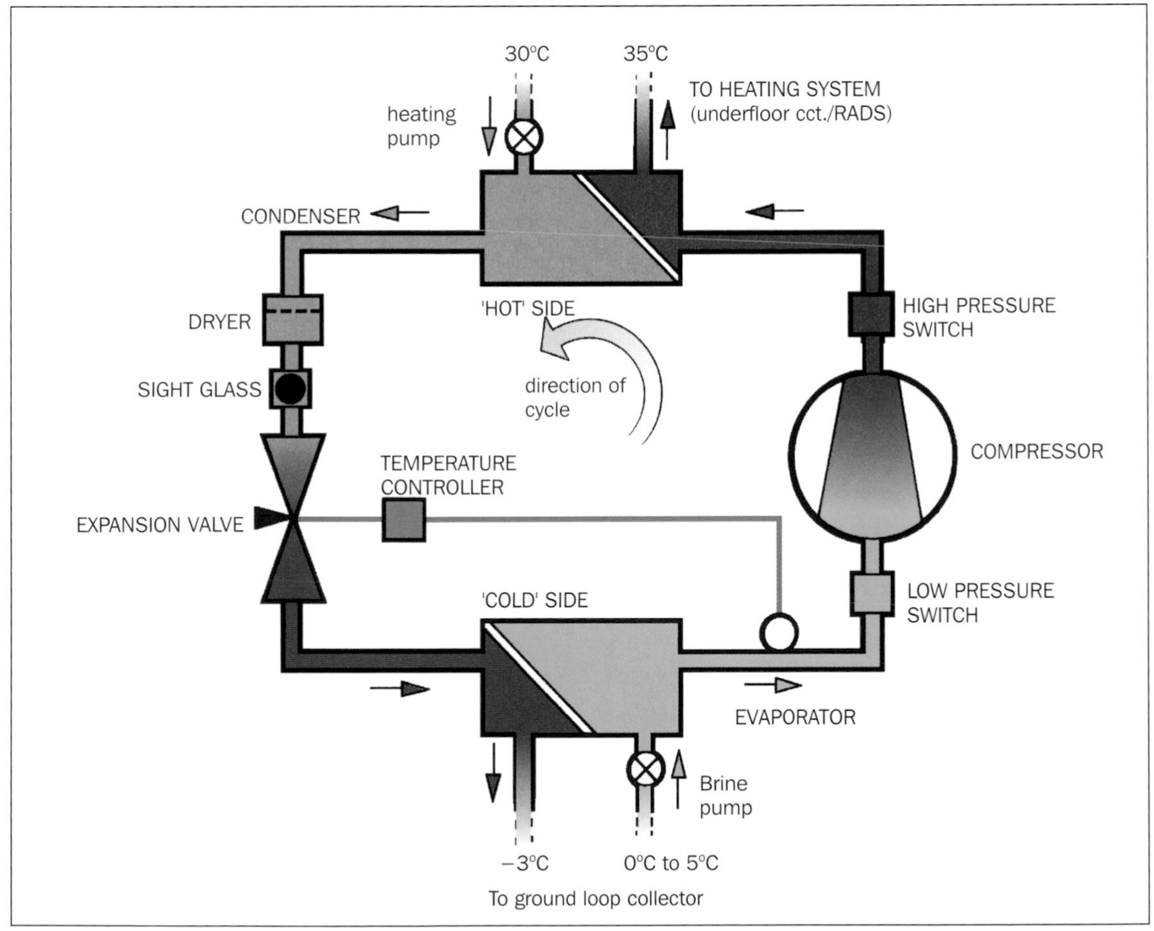

Figure 27: Components of a ground source heat pump.

will lead to loss of refrigerant. A high-pressure switch prevents excessive pressure build-up in the pressurized side of the heat pump to protect against vessel rupture or relief valve opening. Heat pumps are also fitted with drier-receivers to remove moisture ingress, which can corrode and damage the refrigerant system. In serious cases water ice can form, which can block the expansion valve. Heat pumps are also fitted with a sight valve to allow inspection for gas bubbles. These can indicate that the refrigerant quantity is low or that the refrigerant is not totally condensed. It is worth stressing that work on the pressurized parts of the heat pump is specialized and must be carried out by fully qualified refrigeration engineers and this is not normally part of the heat pump installation process. All but the largest industrial heat pumps are hermetically sealed and pressurized, thereby reducing noise, space and heat losses. This means that the compressor and the motor that drives it are encased in a welded metal shell.

AMBIENT HEAT SOURCES FOR HEAT PUMPS

The performance of any heat pump is closely related to the nature and quality of the heat source. Ideally the source would have as high a temperature as possible and the heat would be abundantly available during the heating season. Ground source, air source, water source and exhaust-air source heat pumps are considered useful heat sources for domestic installations. Sea, lake, river water and, to an extent, rock and waste-water heat may also be available to exploit in commercial, public or industrial installations. Heat pumps are installed in a variety of residential, commercial and industrial applications, with the greatest number as reversible air-to-air heat pumps in shops and commercial units.

Ground Source Heat Pumps (GSHPs)

GSHPs are particularly suitable in rural areas where there is greater land availability. Horizontal GSHP collectors require extensive ground areas that are free of large boulders, with a soil depth of around 1.5m, and unshaded by trees or buildings to allow the full strength of the Sun's heat to permeate onto the collector area.

GSHPs are common in both residential and commercial buildings. The ground temperature at a depth of one to two metres is generally in the range 5°C to 13°C in the UK and Ireland. The heat is extracted from the ground using coils of pipe laid horizontally or vertically in the soil. GSHPs can be either direct or indirect systems and these are described below. Their effectiveness very much depends on how damp the soil is and how effectively the heat is replenished through rainfall – especially when the heat is most needed, during the heating season.

Air Source Heat Pumps (ASHPs)

Air source heat pumps, despite their huge popularity in north America and parts of Europe, are only now appearing in the UK and Ireland, where the ambient air temperature very rarely drops below the freezing point of water; it is mostly in the range 0 to 15°C. However, the seasonal performance factor (SPF) is much worse, by 10 to 30 per cent, than that for a ground source heat pump, largely due to the rapid fall-off in performance as air temperature drops (from season to season, at night or in windy conditions). Air source heat pumps are cheaper to install and they do not require the ground or water heat transfer loops, with their associated equipment. They are used in a range of applications such as cooling for lofts, restaurant kitchens, hotels and computer plant rooms, where the hot water produced can be utilized in a range of other ways. Resembling air-conditioning boxes, they can be free-standing, or fixed to the outside walls of buildings or in roof spaces. Recent developments have produced advanced ASHP systems that are capable of providing higher temperatures. These systems are very suitable for the retrofit market. They can reduce the carbon emissions of domestic heating systems, as well as having lower running costs than conventional systems.

Ground Water Heat Pumps (GWHPs)

Ground water heat pumps use water flowing in the ground beneath the surface (referred to as an aquifer), which occurs at temperatures in the range 4 to 10°C. GWHPs are classed as open systems when they pump water from the ground, through the heat pump, and return it back to the ground, albeit not in the same place. A closed GWHP system will operate similarly to the GSHP but with the coils inserted in the ground water source. Open system heat pumps can have problems with freezing, corrosion and fouling, and the installation of the heat transfer loop can be expensive. When considering a ground water heat pump, care needs to be taken that the ground water table is not disturbed or contaminated. GWHPs generally require authorization from the local water provider and this can be an expensive and protracted process.

Exhaust-Air Heat Pumps

Exhaust-air heat pumps (or ventilation heat pumps) are common in various parts of the world, but not in the British Isles. Exhaust-air heat pumps recover the heat that would otherwise be wasted by the building ventilation system. Some of this heat can be returned to the building or to the hot water system. In large buildings exhaust-air heat pumps are often used in combination with air-to-air heat recovery units. This type of heat pump requires continuous operation of the ventilation system, especially during the heating season.

With changes to the UK Building Control Regulations, especially in regard to ventilation, air-tightness and the importance of ventilation control, exhaust-air heat pumps may become more popular. Regulations call for full air changes every two hours, resulting in significant quantities of hot air being removed from homes and other buildings by ventilation. Exhaust-air heating (and cooling) aims to replenish the fresh air in a building, but at the same time recover any heat from the exhaust air.

River or Lake Water Source Heat Pumps (WSHPs)

River or lake water source heat pumps can draw heat from large quantities of water, although the temperature of this water may drop significantly in winter: lakes, for example, can freeze. Seasonal variations in the water temperature render these less efficient than other types of heat pumps. Protection of the associated pipework and access to the source of energy can present problems for this particular design.

Rock Heat Pumps

Rock heat pumps are used where there is poor availability of ground water, but available heat in the underlying rock. They require boreholes sunk to depths of 100 to 200 metres, and this is an expensive operation. These systems require a detailed geotechnical survey, which may be costly if not readily available.

Seawater Heat Pumps

Seawater heat pumps are generally employed in large applications, where a significant heat source is required. At a depth of 25 to 50 metres, the temperature of seawater is constant (5–8°C). It is important to use corrosion-resistant pipework and to give consideration to the organic fouling that can result. It is also essential to protect against the tidal movement of rocks and damage from anchoring arrangements for ships and boats.

Waste-Water and Effluent Heat Pumps

Waste-water and effluent heat pumps can benefit from relatively high waste stream temperatures available throughout the year. The application would mainly be in the water management and industrial sectors, where significant energy savings are possible.

HEAT PUMP SYSTEM DESIGN

The basic component parts of a heat pump system are:

- The external heat collector, which takes heat from the ground, water or air.
- The pipework and circulating pump between the collector and the heat pump.
- The heat pump.
- A buffer hot water storage tank, which is an option for wet heating systems.
- Pipework or ductwork connection for air ventilation heating systems (where used).
- Space heating system, including circulation pump, radiators or upstairs and downstairs under-floor heating array.

Heat pumps are a proven, safe and environmentally friendly alternative to fossil fuels and they are cost effective in suitable domestic and commercial applications. They can also provide a heating option where mains gas is not available. They have the following additional advantages when compared with conventional space and water heating systems:

- Heat pumps have high reliability: they have few moving parts and no external components exposed to the weather.
- They have long life expectancy: up to twenty-five years for the machinery and up to fifty years for the ground loop.
- They generate very little noise.
- They are safe close to animals or pets.
- They can be located almost anywhere within a building, acknowledging that there may be some vibration and that pipe runs should be as short as possible.
- They do not have a fuel store, although most systems should include a hot water buffer tank.
- They do not have explosive gases or use combustion as there is no boiler.
- They do not need a flue or have ventilation requirements.
- They are non-polluting.
- The heat pump cycle can be reversed to provide space cooling, although this option must be exercised with care since the pump will have been optimized to perform in heating mode, and modifications to this will degrade performance.

The golden rules for heat pump systems are: use good quality under-floor heating upstairs and downstairs, super-insulate the building, and make sure the heat load calculation for the property is performed, so that the heat pump is correctly sized. It is essential to ensure that the ground conditions are suitable for the expected heat uptake from the pump and it is helpful if the local energy supply company can provide an off-peak electricity tariff; some may even provide a special heat pump tariff.

The heat provided from the source is largely free, but there is a cost associated with moving it round the system, so heat pumps are not entirely without energy cost. The cost arises from the heat pump machinery, its installation and the electrical energy to run it. If a heating system is changed from oil heating to a heat pump, the overall energy cost will be lower: the oil bill will fall to zero, but the electricity bill will increase because the heat pump requires electricity to provide some of the heat (not just to run the pumps).

Heat pumps are referred to as air-to-air, where the ambient air is the heat source and hot air is ventilated to provide space heating, or air-to-water, where ambient air is again the heat source and hot water is circulated to provide space heating. Similarly they are also referred to as water-to-air and water-to-water where the heating sources are water based and the space heating is delivered through air and water respectively.

The ideal stage to install a heat pump, particularly a GSHP, is when the building is being constructed; a prerequisite is that sufficient land area must be available to provide the necessary heat source. The installation of the ground works is best completed at the construction stage when the diggers are on site. Recommendations for the dimensions of trenches or boreholes will vary between heat pump manufacturers. The ground above the collector loop can be fully utilized and a rule of thumb suggests that at least two-and-a-half times the footprint area of the house needs to be committed to ground loop pipework. Greater area should be used if it is available, but obviously costs will rise as the area increases because more digging and additional piping will be required and also a bigger collector circulating pump will be needed. If there are existing water sources (lakes and ponds) available, or where it is proposed to install Sustainable Urban Drainage Systems (SUDS), the opportunity to install WSHPs within the body of the water should be considered. Similarly in larger developments, such as schools or community buildings, then GSHP ground loops could be laid beneath green spaces or playing fields.

Vertical Boreholes

Vertical boreholes or vertical collectors are the more expensive option, but they can be used when there is insufficient land area available or the ground conditions are unsuitable. The vertical collectors are inserted into pre-drilled boreholes, which are around 15 to 120 metres deep and of 100 to 150mm diameter. Boreholes should be a minimum of 6 metres apart. Boreholes give a high thermal efficiency and require less pipe and lower pumping energy. Larger residential buildings or schools may require a number of boreholes.

Electrical Requirements

When the electric motor that drives the heat pump compressor starts up, it can cause disturbance on the electricity network. Some motors draw a large start-up current, which can cause lights to flicker or the supply voltage to dip and surge (this is known as a spike). This spike can be particularly severe on buildings served by a single-phase electricity supply. Before connecting a heat pump, it is advisable for the installer to contact the distribution network operator (DNO, previously known as the electricity company), since this problem may limit the capacity (kW) and design of heat pump that can be used. Heat pumps fitted with soft-start controls or that have been designed with low start-up torque compressors will be less affected. This is not such an issue in mainland European electricity systems, because three-phase electricity supplies are frequently provided to houses. However, this means that many European-supplied heat pumps are only suited to work with three-phase supplies. Single-phase supplies can generally only carry heat pumps up to a rating of 19.5kW.

Open Loop and Closed Loop Ground Source Heat Pumps

Open loop GSHPs bring water from under the ground to the surface, extract the heat and return the colder water to the ground. They do not have a system of pipes in the ground to act as a heat collector and instead rely on the ground water. Until recently open loop heat exchangers were the most widely used in Canada, America and Europe, especially in commercial applications. These systems can be very cost-effective where a source of groundwater is available, because water can be delivered and returned using relatively inexpensive wells that require little ground area. However, lack of available water, corrosion and water contamination issues mean that the focus has now changed to closed loop systems (or ground coupled systems).

In closed loop systems, the heat collector is a buried loop of sealed pipe with a heat transfer fluid flowing through it. These ground loops of pipe may be horizontally or vertically arrayed. They do not abstract or return water, or any other fluid, from the ground.

DIRECT AND INDIRECT AND INDIRECT HEAT PUMP SYSTEMS

Indirect Heat Pump Systems

As previously mentioned, GSHP heat exchangers may be of two designs, direct and indirect. In the indirect system, which is the most widely available, a sealed loop of high-density polyethylene pipe is used to carry heat transfer fluid into the ground (the collector). The collector is buried in a shallow trench around 1.5 metres deep, or it may be sunk in one or more vertical boreholes. The pipes must be buried at least below the frost level, which for most of the British Isles is around 0.75 metres depth. Typically, horizontal collector loops are installed below a minimum depth of about 1 metre (see below). In trenching, excavation costs are high relative to pipework costs, so some designs employ slinky coils to maximize the heat transfer per unit area. Slinky coils are laid as a series of overlapping pipe coils, placed vertically in a narrow trench, or horizontally at the bottom of a wider trench. Trench lengths may be up to 30 per cent shorter using slinky coils, although pipe lengths could be 200 per cent longer for the equivalent thermal performance. When using slinky-type collectors, care should be taken, due to the high density of the pipework laid in a relatively small area, that the over-extraction or removal of available heat does not lead to sub-soil freezing (known as frost heave). Incorrectly sized heat pumps or poorly laid collectors can freeze the soil.

Indirect systems use high-density polyethylene for its flexibility and also because it can be joined by high temperature fusion (or melt welding). The pipe diameter must be big enough to keep the pumping power low but small enough to ensure that there is sufficient turbulence in the water/brine mixture to get good heat transfer from the ground. Pipes are usually 20 to 40mm in diameter.

Direct Heat Pump Systems (DX Systems)

In direct systems the refrigerant is circulated directly through the ground loop collector, which is normally made of copper. This is referred to as a direct expansion system or DX system. This mode of operation

Figure 28: The 11kW NIBE Fighter 2005 air source heat pump is located on the ground external to the building. (Author/www.nibe.co.uk)

Figure 29: The only component parts visible inside the building are the fully enclosed heat pump (20kW TERRA) (about the same size as a medium-sized fridge), the circulating pumps and under-floor heating manifolds. (Author/www.idm-energie.com)

will give greater efficiency than an indirect system because the ground loop heat exchanger has been eliminated and there is no additional ground loop circulation pump. This also means that a shorter ground loop is required, saving material costs, although more refrigerant is required and the potential for loss of refrigerant with associated high repair costs is greater. In direct expansion systems, copper pipe of 12 to 15mm diameter is used in the collector and it must be coated to prevent corrosion.

DX systems are generally regarded as most suited for smaller domestic installations since only one heat exchanger is involved and the required collector area is less. DX systems can only be installed horizontally.

Air Source Heat Pumps and Bivalent Operation

ASHPs operate best when the air temperature does not drop below freezing. They have not traditionally been used widely in northern Europe due to the freezing winter air temperatures, but they are used extensively in Canada and the USA. The effectiveness of an air source heat pump will decrease quickly with falling air temperature. On a cold day, ground and water source heat pumps are more resilient since the temperature of the air falls much quicker than that of the soil or water. While air source heat pumps have lower capital costs, the air temperature is lowest when the heat is required most. This means that in most instances the heating requirements of a building can never be entirely dependent on the heat output from an air source heat pump, thereby requiring an auxiliary heat source.

Heat pumps that satisfy the space heating and hot water requirements without the back-up of an auxiliary source are known as monovalent. If an ASHP has an auxiliary heating source it is known as a bivalent system. ASHPs have advanced technically in recent years and they are becoming more numerous. In the past there has been resistance to fitting them, primarily because they do not have a long operating history in this climate and also they require a back-up heating source, usually fossil fuel.

Heat pumps can be retrofitted to sufficiently insulated existing properties but they will perform much better in a new building. New build avoids being forced to use existing radiators and insulation tends to be inadequate in existing houses.

HEAT PUMP SPACE HEATING SYSTEMS

Renewable energy space heating systems such as wood fuel and heat pumps can offer a competitive alternative to fossil fuels and they are normally sized to provide space heating plus hot water. Heat pumps can also provide cooling, but in our northern latitudes this is unlikely to be necessary in domestic properties, neither is it likely to be utilized very often. Operating a heat pump in cooling mode will lead to higher electricity consumption.

Although heat pumps can be used to directly replace existing fossil-fuelled boilers supplying radiators, there are several differences in the way the systems operate, which can lead to difficulties. Wet radiator systems operate with higher temperature water (60–80°C) in the radiator circuit and they will deliver heat into a house relatively quickly. A house heated using an oil-fired boiler can reach comfortable room temperatures within an hour from switching on the system. Heat pumps require a longer period to raise the temperature to a comfortable level and the water circulating in the radiators will generally be between 30 and 40°C. Heat pumps operate most effectively with under-floor systems, but, if this is not possible, it may be necessary to increase the heating surface area of the radiators by some 30 to 40 per cent. If radiators must be used, then they need to be of sufficient surface area to deliver the required amount of heat to each room as determined in the heat load calculation for the property.

In well-designed systems, under-floor heating will provide room temperatures to satisfy the occupants' needs and in line with current Building Control Regulation design, which stipulates 25°C maximum.

Buffer Tank

Including a buffer tank in the system will ensure good quantities of hot water and it also means that the heat pump can operate for longer periods without shutting down and starting up. Excessive start-stop cycles reduce the system efficiency and result in wear on the compressor. Compressors are designed to survive a limited number of cycles over their lifetime and heat pumps are generally fitted with restart delay timers to prevent heat pump on–off cycling. Manufacturers design the control software to prevent pump restarts for at least five to ten minutes after a shutdown and this prevents overheating of the motor windings.

Most systems will benefit from a hot water buffer storage tank with a capacity of between 60 and 200 litres, depending on the size of the system, together with heavy insulation of at least 50mm to prevent heat loss. Connecting a solar water heater into the buffer tank through a heat exchange coil will preheat the stored water, raising its temperature and reducing the work required from the heat pump. The capacity of the buffer should be such that the tank can be heated fully during the overnight, lower-price electricity tariff period.

If the system is to be used in conjunction with radiators, then a buffer tank is usually essential. If a variable speed drive is fitted to the compressor or a digital compressor is employed, then the buffer tank can be smaller or removed totally. A buffer tank is not always necessary with under-floor heating systems due to their high thermal inertia, which calls for less frequent start-ups. Where an air-to-air heat pump requires heat from the system to defrost the evaporator, then a buffer tank is included to provide this. Buffer tanks are normally included when the heat pump is used to supply domestic hot water.

Domestic Water Heating from Heat Pumps

Domestic hot water is supplied at a temperature of around 40 to 50°C. If a heat pump is required to deliver water at this temperature, then it will be operating inefficiently and expensively (at a poor CoP). One solution is to use a well-insulated buffer tank, which can be supplemented by an auxiliary heating system such as an immersion heater. As noted in Chaper 2, there is a daily requirement to raise the temperature of the water in a hot water storage cylinder above 60°C to prevent the growth of Legionella bacteria. The efficiency of heat pumps falls as the output water temperature rises, so it is better to use an auxiliary heater to boost heat the domestic hot water to temperatures above 40°C.

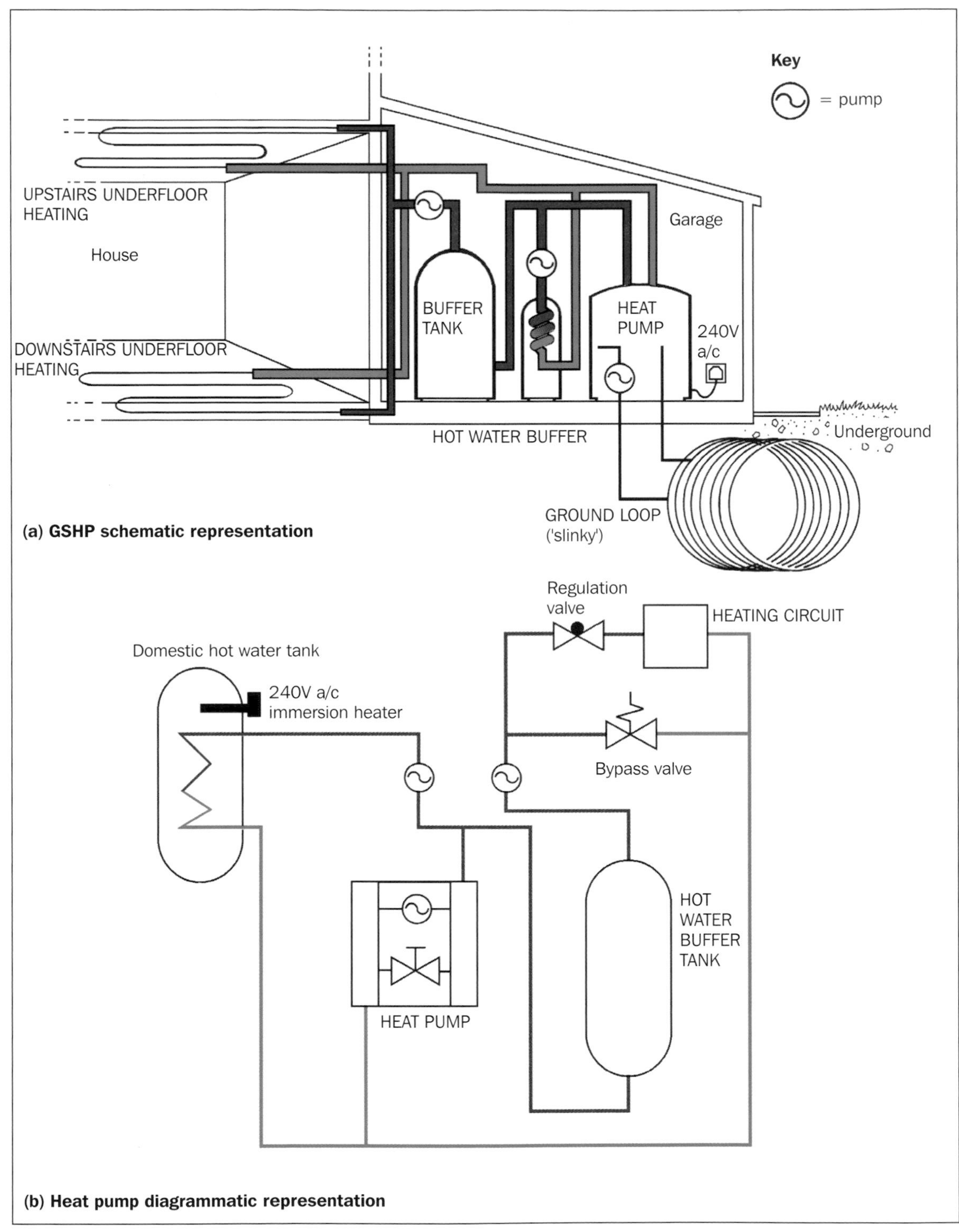

(a) GSHP schematic representation

(b) Heat pump diagrammatic representation

Figure 30: Heat pump installation schematics.

Heat pumps frequently include well-insulated, pre-heat tanks to provide domestic hot water. A pre-heat tank can be fed through an indirect coil from the heat pump space-heating circuit, with boost from an immersion heater.

Some heat pumps are fitted with a desuperheater, which is designed to provide partial domestic hot water. A desuperheater can produce limited volumes of water at a much higher temperature, say 70°C. Since these devices only work when the heat pump is running, it may be more economic to install an immersion heater anyway.

Under-Floor Heating Systems

Under-floor heating systems can provide comfortable room temperatures from a low-temperature hot water source, such as a heat pump. As the name implies, an under-floor system radiates heat from the floor upwards, which provides a more comfortable and even heat distribution in the room. Conventional radiator heating systems provide the majority of their heat through convected air, which rises straight to the ceiling and does not dissipate well into the room. With under-floor heating, however, the room thermostat can be set 2 to 3 degrees lower.

An under-floor heating system consists of a pipework array set on top of the floor surface, which might be concrete or timber. It is good practice to place a thick, insulating mat on top of the bare floor and then to put the heating pipe array on top of the mat. The pipes, once laid in a dry mortar screed and arranged in a pattern to ensure equal heat dispersion, are then sheeted over with the flooring material. Each room in the building, including those upstairs, should have a separate heating circuit. The pipes in an under-floor heating system are between 16 and 25mm in diameter and spacing between them can vary from 80 to 350mm. They are usually made from temperature and pressure rated (110°C, 10 Bar) high-density polyethylene, polybutylene or multi-layer metal-plastic composites. It is important that polyethylene pipe material includes an oxygen diffusion barrier to reduce

Figure 31: An under-floor heating system is the ideal complementary heating system for a heat pump and will allow maximum comfort levels for water temperatures between 30°C and 40°C. (Image courtesy John Ross, South Eastern Regional College, Lisburn, Northern Ireland)

the rate of oxygen transfer, which can cause corrosion of the metallic components of the system.

Under-floor heating systems are designed to have a high thermal inertia and as such they cannot increase in temperature rapidly, so it takes longer to bring a building up to a comfortable temperature. The systems, which include the floor materials as part of the thermal mass, hold the heat and radiate it into the building. Compared to a radiator system, which may take around an hour to heat up a building, an under-floor system will take several hours to achieve the same level of warmth. This means that on-off periods or interruptions in electricity supply tend to have a less dramatic impact on the heat flow from heat pump systems. The European Standard EN 1264-3 for heat pumps stipulates that the maximum temperature of the floor surface must not exceed 29°C in the living area. In other parts of the floor, for example areas with a high heat loss in proximity to external walls, or close to doors or windows, the temperature is permitted to increase to 35°C. Higher temperatures can be achieved in these areas using water at the same flow temperature, but passing through tubes that are spaced closer together. Under-floor heating systems are well suited to continuously circulate hot water to minimize on-off cycles for the heat pump and to eliminate the long response times.

Good practice in the design and installation of under-floor heating systems would include the following:

- The temperature difference between the inlet and outlet water flows from the floor pipework should not exceed 5°C.
- The flooring materials should have good thermal conductivity (to get the heat from the under-floor system out into the room).
- The length of any single pipe in the system should not exceed 100m.
- Each room should have its own heating circuit.
- All heating circuits should be connected to a manifold for flow balancing and temperature control.
- Each circuit should be fitted with an air release vent (so that the system can be properly bled).
- Commissioning and initial heating of the floor should be completed according to the technical guidelines supplied by the manufacturer.
- Fitting an insulation mat below the under-floor pipework is essential to reduce heat loss downwards and to enhance heat-up times.

Once installed, under-floor heating systems are sensitive to the decorative covering. Carpet and wood flooring materials are insulators and will prevent the heat transferring to the room. The positioning of furniture and floor coverings could also prevent heat transfer. It is possible to use wood covering on under-floor heating systems, but the wood and floor adhesives will dry out and may crack. It is prudent to check the suitability of any proposed floor covings with the under-floor heating supplier or manufacturer.

Wall Heating Systems

Currently wall heating systems are uncommon in the UK and Ireland but they can work well, especially if combined with, or if installed instead of, under-floor heating. Wall heating can be appropriate when it is impractical to install under-floor heating, particularly in retrofit installations. Wall heating gives a warm feeling to a room and is particularly suited to lower temperature heating systems.

Wall heating systems are available in two designs. The first utilizes the thermal mass of the wall, and the heating pipe array is integral with the outer plaster finish. The heating loops are fitted straight on to the internal rough wall or brick surface. They are

Figure 32: Wall heating systems are not common in the British Isles, but they can work very effectively with a heat pump. (Image courtesy John Ross, South Eastern Regional College, Lisburn, Northern Ireland)

then covered with a special plaster, with a high heat conductivity and an elasticity that can withstand the expansion and contraction of the pipes as they heat and cool. In this case the wall is used for heat storage and the thermal mass becomes part of the heating system. This type of wall heating is most suited for heating rooms that are in permanent use or for walls between rooms in living areas.

The second type of wall heating system is where the heating array is applied on top of a layer of insulation covering the internal rough wall; the heating system is thus de-coupled from the thermal mass of the wall, which cannot be used for heat storage. In this case the room heats up and cools down more rapidly.

In both cases it is essential that the position of furniture, wall decoration and pictures does not obstruct the heating area of the walls. Extra care should be taken when hanging pictures or wall decorations on the walls since nails could damage the heating array. Heavy decorations will impede heat flow to the room.

Good practice suggests that when using wall heating systems the temperature should be in the range 25 to 40°C with a maximum of 45°C. As with under-floor systems, the temperature difference between supply and return should be no more than 5°C. Each room should have its own separate circuit, complete with air vent, connected to a common manifold. It is recommended that at least 10mm of reinforced plaster covers the pipes at their highest point of protrusion from the wall.

Using Radiators with Heat Pumps

Existing buildings across Europe are mostly heated by wet radiator systems. Plumbers and designers are familiar with radiators and their application, insofar as they are adaptable in terms of size, design and position within a room; due to their widespread use they are also inexpensive. They require high water temperatures, however, which make them unsuitable for retrofit heat pump systems. If they are to be used with a lower water temperature system, then radiators with a much higher surface area will be required and this may make them unsightly and aesthetically unsatisfactory. The quality of heat produced and the heat distribution pattern from radiators is not as good as that from wall or under-floor heating.

SIZING A HEAT PUMP TO MATCH A BUILDING'S HEAT LOAD

A heat pump's size in relation to the building it is to heat is critical to the successful operation of the system. If the pump is too small or too big for the building's heat demand, it will use excessive amounts of electricity wastefully and high electricity bills will make the system uneconomic. It is also of paramount importance that a heat pump will have sufficient temperature pick-up from the heat source to meet the building's heat demand over the coldest period.

Over-sizing the heat pump has a number of consequences, including inefficient performance and reduced compressor lifetime, not to mention the significant increase in capital and running cost. Installing an oversized heat pump will result in continuous on-off cycling, giving inefficient operation with poor CoP and SPF. Increasing a heat pump's capacity, even by a relatively small amount, can result in a much higher capital cost. Under-sizing will mean that the shortfall in heat requirement will need to be met from an auxiliary heating source, usually electric heating using immersion heaters, which lead to excessive electricity costs.

European Standards EN 15450 and EN 12055 B/0W35 recommend that it is good practice to size the heat pump for total space heating and domestic hot water supply. An industry rule of thumb suggests that domestic hot water can be allowed for if the heat pump capacity is increased by around 0.25kW per occupant.

Heat pumps are typically sized to provide 100 per cent of the peak load on the coldest expected day. Due to the relatively few days in the year when this will be experienced, the vast majority, perhaps 95 per cent of the heat load for the building will be delivered by the heat pump. The balance must be supplied by an auxiliary source, usually one or more immersion heaters. Rating the heat pump just below the peak heat load means that, for the majority of the year, the heat pump will trickle heat continuously to keep the building at the necessary temperature. If the heat pump is to run to suit the low cost electricity tariff periods, then the capacity must be adjusted to allow for the reduced hours of operation.

Figure 33: A vertical bore-drilling rig can drill down several hundred metres. Multiple boreholes can be sunk to provide more heat.

To achieve maximum efficiency from the heat pump, the supply temperature to the building should be kept as low as possible, setting 40°C as a target; it should certainly be below 50°C. SPF and CoP from the heat pump falls off rapidly with higher supply temperatures.

For existing properties with radiators fitted, it is common practice to estimate a heat load for each room based on the radiators already installed. This is not good practice since the original heat loss calculation may not have been carried out correctly, if at all, or the property may have been subsequently modified. The golden rule is always to perform an accurate heat loss calculation for the building. This exercise will incur an additional cost and should be allowed for in the pricing.

The integration of another renewable energy source to provide hot water or to pre-heat feed water could be considered, particularly where grants are available. Solar hot water panels would be suitable for providing hot water or as pre-heat for space heating. A renewable source of electricity such as a wind turbine or solar photovoltaic panels will help offset the electrical running cost of the heat pump.

GSHPS AND GROUND CONDITIONS

For a GSHP, soil conditions are critical. Rocky or well-drained soils are not the best ground conditions in which a heat pump can operate. Damp or marshy ground conditions are better, since these allow the heat to be replenished through rainfall or ground water flow around the collector. If the heat pump has been properly sized to suit the building it is to heat, a rule of thumb suggests that a heat collector area of at least two-and-a-half times the ground area of the building is required. If there is insufficient heat available in the ground to provide the required amount of heat, the pump can draw so much heat from the ground that it will freeze solid.

The ground temperature at around 1.5 metres below the surface remains relatively constant throughout the year, which means that predictions of the heat available in the winter can be made with some confidence. A cheaper option is to place heat transfer loops in water if a suitable resource (lake, river or pond) is available. If space is restricted or ground conditions are not favourable, then another option is to sink a vertical borehole, although this would prove to be more costly. Vertical boreholes can either be indirect or direct heat transfer systems.

Heat pumps, more so than conventional heating systems, should only be operated within highly insulated buildings. Fully insulated properties with a correctly sized heat pump will benefit significantly in terms of energy, efficiency and economy.

Sizing of the ground loop is complex and critical to the performance of the heat pump system, and a number of software tools are available to assist a designer in getting it right. Ground loop costs account for around 30 to 50 per cent of the overall system costs, so over-sizing will be uneconomic. Under-sizing would be unsatisfactory, since the heat pump would be unable to deliver the necessary energy to satisfy the heating requirement of the building. It is also very important to consider the rate at which the heat will be drawn from the ground and how quickly it can be replenished. A list of relevant calculations that can give a good estimate of the size of the ground loop and the heat pump output required, for a given set of conditions, is available from the IEA Heat Pump Centre (for contact details see page 75).

The circulating fluid in the ground loop has a freezing point of around -10°C, which is well below the mean heat pump outlet to the ground loop temperature, which is typically -3°C. The pump's safety protection (low temperature cut-out) from frost damage is normally set at –5°C. It is important to remember that antifreeze–water mixtures become more viscous as temperature falls and this will result in a higher collector loop pumping load. As a general rule, a ground loop circulating pump that can tolerate low temperatures with a rating of around 50W per kW of installed heat pump capacity is used: for a 20kW heat pump the pump rating will be 1kW.

HORIZONTAL 'VERSUS' VERTICAL COLLECTORS FOR GSHPS

Installing horizontal collector arrays is relatively simple and inexpensive since they require low-cost equipment. Relative to a borehole, horizontal arrays will need longer lengths of collector pipe to allow for seasonal temperature and ground moisture content variations and therefore a larger collector area. Where land area is constrained, vertical installations or a compact slinky horizontal installation may be suitable. If local conditions include hard rock, then a vertical installation may be the only available choice. The vertical option is more expensive since drilling costs are high, although the heat collector is deeper in the ground, reaching higher temperatures. Drilling a borehole could take up to two or three days, depending on the consistency of the ground.

Table 12: Ground conditions classification.

Ground Type	Defining Characteristics	Classification
Dry non-cohesive soils	Crumbles immediately, cannot be formed into a ball (e.g. sands and gravels).	Poor, low heat capacity soils.
Cohesive soils	Does not crumble. Binds well and can be rolled into a ball (clay, loam, boulder clays).	Good, reasonable heat capacity available.
Water-saturated sand/gravel	Sand or gravel that is permanently saturated with water.	Excellent, good heat capacity available.

Example of Heat Pump Sizing

A heat pump ground loop collector is to be located in an un-shaded area of grassland roughly 15 metres by 22 metres (330m^2). The heat load of the super-insulated building has been assessed as 7kW. The heat pump is unavailable twice a day for periods of one hour.

The availability of the heat pump = 24 – (1 × 2) = 22 hours

The heating power of the heat pump is given as:

= (building heat load) × (24 hours)/availability, thus,

the heating power = 7 × 24/22 = 7.6kW

The ground loop inlet temperature is 5°C (to the heat pump) and the outlet temperature to the under-floor heating is 35°C.

The heat pump will be selected from a range of manufacturers' specifications to be:

Effective heating power: 8.1kW

Electrical power consumption (pumps and compressor): 1.93kW

Figure 34: The ground loop collector pipework is installed in a series of trenches (T) and the individual loops are returned to a connecting manifold. The final connections (flow and return, A and B respectively) to the building are indicated.

So the capacity of the heat pump is: 6.17kW

The 330m^2 area of soil is classed as cohesive damp and is estimated to provide heat at around 20W/m^2 of pipe laid in the grassland. This gives a heat capacity for the collector of 330 × 20 = 6.6kW. Since the required capacity of the GSHP is 6.17kW then this area of ground should provide enough heat to satisfy the maximum heat demand of the house.

A further calculation can be performed to estimate the sizing of the collector pipework. The heat pump capacity is 6.17kW and the ground capacity is 20W/m^2, so the collector area required is 308.5m^2. If the spacing between pipes is 80cm, then the pipe length required is 308.5/0.8 = 385.6m of pipe – rounding up will give four collection loops of 100m each. In fact with 400m of pipe at 20W/m length, the potential heat available should be 8kW, which represents a good margin of comfort for operating the pump.

Figure 35: Close-up view of the manifold showing the individual ground loop collector circuits.

PLANNING AND BUILDING CONTROL REQUIREMENTS FOR HEAT PUMPS

Heat pumps generally do not require planning permission, indeed under some interpretations of the planning requirements they are considered as permitted development, so long as all of the equipment is included within the building. As with other renewable energy technologies, it is advisable to check with the planning office, especially if the building is listed or if it is within a conservation area. If the heat pump is to have a large buffer tank that is to be housed in a new or separate building, then it would be prudent to check with the planning office. Air source heat pumps can be mounted on the outside walls of buildings or on the ground, and if these are listed buildings or in a conservation area, then consideration will be needed to ensure they are sensitively sited. Under the Building Control Regulations, where heat pumps may be defined as controlled heat sources, they can be used in a trade-off situation to reach approval under SAP rating calculations.

As the installation of GSHPs will require the excavation of trenches or deep boreholes, it is important to consider in advance the possibility that archaeological remains could be on the site. In certain regions the construction or extraction of a borehole or well for the purpose of abstraction, or the extraction or discharge of water to the environment, may require authorization from the relevant local authority. Similar authorization may be required from the relevant water control authorities when a WSHP is being considered and the input heat source loop is to be submerged in a lake or river.

ENVIRONMENTAL IMPACT OF HEAT PUMPS

The electrical energy used by a heat pump and the heat energy drawn from the ambient heat source replaces the fossil fuel energy consumed in a conventional boiler heating system. Electricity bills will increase once the heat pump starts operating, but the coal, oil or gas bills will disappear. The electricity consumed and the size of the bill is crucially dependent on matching the heat pump to the heat load (and heat source) correctly. If it is too big then the pump runs inefficiently with frequent starting and stopping; too small and the pump runs continually, never able to keep up with the demand for heat. Any environmental evaluation of a heat pump application needs to account for the indirect emissions related to the generation of electricity used to operate the heat pump. Many people look to offset the carbon dioxide released by the fossil fuel-generated electricity used to run the heat pump by signing up to a green electricity tariff. Better still, some install PV modules or a wind turbine to ensure the electricity consumed is both renewable and generated on site.

The environmental footprint of a heat pump is much better than that of a conventional heating system, with much lower local emissions. However, heat pumps may contribute to greenhouse gas emissions through the leakage of refrigerant over their life cycle. In addition to leakage, which happens during operation, the decommissioning of the heat pump will require the removal and disposal of the refrigerant. The most commonly used refrigerants have been the hydrofluorocarbons (HFCs). These gases have no ozone depletion potential, but they do contribute to global warming. HFCs are controlled under a European Directive in an attempt to improve losses during heat pump operation, inspection and recovery.

Heat pumps are generally considered very safe and quiet with little exposed equipment, particularly GSHPs. Some minor vibration may be experienced, although this can be minimized with anti-vibration pads or rubber mats. They have no open flames, flammable fuel or potentially dangerous fuel storage requirements.

ECONOMICS OF HEAT PUMPS

Heat pumps operate with high efficiencies and they can provide energy and emissions savings. Installation costs are higher than for conventional heating systems, but this is offset by reduced operating and maintenance costs. There is the additional benefit of security of supply, since the fuel is fully available locally and not controlled externally. The costs of the heat pump system are largely determined by the installed capacity, but the ground conditions and the energy demand from the building can also have a bearing.

In the UK the uptake of heat pumps has been encouraged through the introduction of capital grants. Grant support for heat pumps has also been available in the Republic of Ireland. Due to the difficulty in estimating uptake in these schemes, the duration and level of available grant funding is subject to change at any time. (For contact details for these schemes, their accredited heat pumps and approved installers, *see* pages 25–27.)

The running costs for the heat pump depend entirely on the electricity tariff and this will vary for each energy supply company. Most companies operate an off-peak tariff, when the charge for electricity consumed away from peak times (late evening and overnight) is less than the normal rate; indeed some companies operate a special heat pump tariff. The bottom line for cost comparisons with a fossil-fuelled central heating system is the annual electricity cost attributed to the heat pump operation, against the annual costs for the coal, oil or gas.

The maintenance costs for heat pumps are minimal since there is no requirement for an annual safety inspection, as would be the case with boilers. The collector loop circulation pumps do occasionally fail, although generally they last for many years. Scroll compressors have design lifetimes of at least twenty-five years, but for some other types this might be closer to twenty years. The refrigerant circuit is sealed for life and it is only in unusual circumstances that this part of the system will require attention, but any work must be carried out by a refrigeration specialist. For GSHPs the ground loops have long operational lives: over thirty years for a DX system copper loop and in excess of fifty years for polyethylene pipework.

Table 13: Cost of heat pump principal components.

Component	Potential Typical Cost (%) (Including Installation)
Design, equipment mobilization and commissioning	20–30
Ground loop installation (including trenching excavation)	30–35
Borehole (if required)	50–60
Heat pump system	50–60
Associated equipment and pipework	5–10
Electrical (upgrade of connection, if required)	5–10
Buffer tank (if included)	< 5

Table 14: Typical heat pump installed costs, based on actual fully installed costs.

Technology	Approx. Installation Size	Average Approx. Cost (2008)
Air source heat pump	12kW	£7,700
Ground source heat pump	8–10kW	£7,500
	11–12kW	£8,400
	13–16kW	£9,500
	16–20kW	£10,900
Water source heat pump	16kW	£10,350

Source: LCBP, Reconnect and Greener Homes costs of installed systems

Table 15: Golden rules for heat pumps.	
Top Tips for Heat Pumps	**Things to Avoid with Heat Pumps**
The ideal situation is to include a heat pump as an integral part of a new building. This will allow the system to be optimized for the property and the ground conditions. The building should be super-insulated (new or existing), above the required Building Control standards.	Do not estimate the heat pump size using rules of thumb. Make sure the installer/architect completes an accurate heat loss calculation (retain this information for safe keeping).
Check out the soil conditions for suitability if installing a GSHP.	Do not overestimate the heat available from the heat source. Correctly size the heat source collector. Make sure the installer calculates the length of collector pipe required accurately.
Use under-floor or wall heating (upstairs and down) if possible. If forced to use radiator systems, then these may need to be oversized by 30 to 40 per cent (or use double radiators). In under-floor heating, pay attention to the floor covering – heavy natural wood coverings or thick carpets will impede the heat flow into the rooms.	Do not go with the first installer quotation. Always get at least two quotes, and ask for details of previous installations. Seek references and follow them up.
Contact the DNO (electricity company) at an early stage to establish any potential limitations on heat pump capacity or start-up. Use a heat pump with soft start control. A three-phase line connection to the electricity grid will be needed for heat pumps above 19.5kW.	Do not run the heat pump on domestic tariff electricity: ask the local electricity supply company for the off-peak tariff, or better still they may offer a special heat pump tariff.
Use a buffer hot water storage tank. Ensure that the electrical installation work for the heat pump is completed by a competent, suitably qualified electrician.	Do not use rigid connectors (use flexible couplings) to make the final connections to the heat pump (this avoids noise and vibration). For example, use a rubber mat to set the heat pump on for tiled floors. Do not use couplings to join or bend pipework in the collector for GSHPs.
Use a green electricity tariff, a wind turbine or PV array to offset electricity costs and emissions.	Do not fit a heat pump in an existing poorly insulated building – it simply will not work.

A good installer will give the occupier a thorough understanding of the heat pump controls and will provide fully marked-up, as-installed drawings. This is especially important for heat pumps, where the location and extent of the collector will not be obvious. The pipe runs for heat pumps are complex and sometimes difficult to follow. Pipes and fluid flow directions should be clearly identified as this will greatly aid future fault finding, should it ever become necessary. It is recommended that an accurate site layout drawing is completed, showing precisely the collector area and orientation in relation to the site buildings, with pipe runs and manifolds clearly identified. Once installed, it is good practice to ensure that the system designer or installer monitors performance during the early years of operation. This would give a picture of how the system is behaving over a full seasonal cycle, paying close attention to the heat source. Fine adjustment of the controls may be needed to ensure efficient running. Also optimization of the heat pump operation for the occupants' usage pattern will further increase effectiveness.

HEAT PUMP GLOSSARY

BTU British Thermal Units (historical method of measuring a quantity of heat). The amount of heat required to raise one pound of water through 1°F. Comparative figure = Watts × 3.413.

CFM Cubic feet per minute of air flow.

Condenser The heat rejecting mechanism in a heat pump. The rejected heat is used as space or water heating.

CoP The coefficient of performance of a heating system is the ratio of the heat abstracted from the ambient heat source, divided by the heat put in to the system electrically.

Degree day The number of degrees that the mean temperature for that day is below 65°F (for example, given a mean temperature of 40°F for the day, then 65 – 40 = 25 degree days). Traditionally Degree day calculations have used degrees Fahrenheit.

Desuperheating Where sufficient heat is removed from the vapour to turn it back into a liquid.

Frost level The depth below ground (m), below which frost does not form under extreme temperature conditions.

EER The Energy Efficiency Ratio is the same as the CoP but expressed in terms of BTUs. CoP = 3.413 × EER.

Evaporator The heat-absorbing mechanism in a heat pump.

Flat plate heat exchangers These permit the transfer of heat from a hot medium to a cooler medium.

HA Housing Association.

Heat pump cycle A set of thermodynamic conditions that define the operations that take place in the heat pumps.

Heat sink The area where heat is deposited or to be used (inside a home, etc.).

Heat source The ambient source from which heat is taken or abstracted (water, air, etc.).

Heating season In the UK and Ireland this is generally the months September through to April.

Heat pump Any device that moves heat from one place to another.

kWh Kilowatt hours.

Low grade heat Heat at a lower temperature than is normally considered useful for space or water heating.

Refrigeration cycle The refrigeration cycle uses a fluid, called a refrigerant, to move heat from one place to another. It works because the refrigerant boils at a much lower temperature than water at the same pressure.

SAP rating An approved method to demonstrate the heat loss and carbon dioxide emissions from a building. The SAP rating is used to fulfil requirements of the Building Regulations to notify and display an energy rating in new dwellings.

SCoP The Seasonal Coefficient of Performance is the average CoP over the entire heating season, equivalent to the Seasonal Performance Factor.

SEER Average EER over the entire cooling season.

Single-phase/three-phase Single-phase supply is the normal domestic 240V AC electricity supply with live and neutral conductions. Three-phase supply has three live conductors and a common neutral, and allows for much higher power transfer.

HEAT PUMP GLOSSARY *continued*

Super-insulated Insulated above the requirements of the Building Control Regulations.

Variable speed drive (VSD) As opposed to a constant speed drive, VSD can provide substantial savings. The load on the compressor will vary and this can be controlled using a variable speed drive motor to provide a greater efficiency of operation.

USEFUL HEAT PUMP CONTACTS AND WEBSITES

Heat Pump Groups

The Federation of Environmental Trade Associations (FETA)
www.feta.co.uk
The Heat Pump Association, Henley Road, Medmenham, Marlow, Bucks SL7 2ER
Tel: 01491 578674 Fax: 01491 575024
Email: info@feta.co.uk

IEA Heat Pump Centre
www.heatpumpcentre.org/
Novem BV, PO Box 17, 6130 AA Sittard, Netherlands
Tel: +31-46-4202236 Fax: +31-46-4510389
Email: hpc@heatpumpcentre.org
Useful for the calculations required when sizing ground loops. For heat pump sizing guidance, contact information and specific design tools see HPC-AR12, *Designing Heat Pump Systems: Users' Experience with Software, Guides and Handbooks.*

International GSHP Association
Oklahoma State University, 499 Cordell South, Stillwater, Oklahoma 74078-8010, USA
Toll-Free: 1-800-626-4747 Tel: (405) 744-5175
Fax: (405) 744-5283

UK Heat Pump Network
www.heatpumpnet.org.uk
NIFES House, Sinderland Road, Broadheath, Altrincham, Cheshire WA14 5HQ
Tel: 0161 928 5791 Fax: 0161 926 8718
E-mail: secretariat@heatpumpnet.org.uk
The Government's Housing Energy Efficiency Best Practice programme provides impartial, authoritative information on energy efficiency techniques and technologies in housing. Website at www.housingenergy.org.uk.

Useful Information

Heat Pumps in the UK: A Monitoring Report, Energy Saving Trust GIR72 (download)

Domestic Ground Source Heat Pumps Design and Installation of Closed-Loop Systems, 2004 Edition, Energy Saving Trust CE82/GPG339 (download)

EN 15450:2007. 24: *Heating Systems in Buildings: Design of Heat Pump Heating Systems*

EN 12055:2007: *Liquid Chilling Packages and Heat Pumps, with Electrically Driven Compressors – Cooling Mode – Definitions, Testing and Requirements*

Useful Websites

For detailed information on heat pump use:

National Energy Foundation: Ground Source Heat Pump Association
www.nef.org.uk/gshp/index.htm

In the Republic of Ireland the Department of Environment (DoE) manages the RoI Building Regulations: www.environ.ie

Ground Source Heat Pump Information Centre
www.virtualpet.com/portals/okenergy/gshp.htm

European Heat Pump Standards and Courses
www.eucert.fiz-karlsruhe.de/

CHAPTER 4

Biofuels

Biofuels are carbon neutral fuels that provide an alternative to coal, gas, oil or LPG, especially in rural areas where fuel choice may be limited. They also provide an additional source of rural income when grown as energy crops. Some biofuels are wastes that can be treated in an anaerobic digester. The products can be used to balance soil nutrients and provide both heat and electricity.

ENERGY FROM WOOD AND WASTE

The term 'biofuels' has come to mean a variety of energy sources ranging from a basic wood fuel, such as logs or wood chips (biomass), to include waste digestion by anaerobic bacteria, which produce biogas. A useful definition might be 'biofuels are solid, liquid, or gaseous fuels produced from organic materials, whether directly from plants or indirectly from industrial, commercial, domestic or agricultural wastes. They can be derived from a wide range of materials and produced in a variety of ways' (Energy Technology Support Unit, 1991). When setting national targets for increased levels of renewable electricity, a contribution from biofuels is usually included. Producing electricity from biofuels could involve large-scale schemes, such as cooperation by a number of farms to provide several megawatts (MW) of electricity from purpose-grown energy crops, or perhaps from community-owned and operated anaerobic digesters. Biofuels are much more likely to provide significant quantities of heat rather than electricity, although as yet there are no specific targets relating to heat produced from renewable sources. Governments have included wastes as a biofuel and this derives from the large proportion of organic materials included in our waste streams, including industrial, commercial and agricultural wastes. The United Kingdom and Ireland are not as highly forested as other parts of Europe or America, so the availability of wood waste is less than in other countries such as Sweden, Finland or Austria. Nevertheless, there is an established wood-processing industry that produces significant quantities of wood waste as off-cuts, sawdust and bark. The production of liquid fuels from crops, mostly from grain or seeds, for use in transport or for the pharmaceutical industry is not considered in detail in this book.

WOOD AS A CARBON-NEUTRAL SUSTAINABLE FUEL

Some people struggle with the concept that biofuels are classed as renewable energy sources, especially since they usually require to be burnt in a boiler. Like most fuels (including coal, oil and gas), biofuels burn in the presence of oxygen to change chemically and release energy (as heat). As is also the case with conventional fuels, they release carbon dioxide (CO_2) during combustion. Biofuels are mostly derived from living or recently dead biomass – that is, complex biological materials, like wood, grasses and other plants, which contain lots of hydrogen (H), carbon (C), a little nitrogen (N_2) and some oxygen (O_2). Biomass will also decompose naturally, or oxidize, to release carbon dioxide and water. Unlike conventional fuels, biofuels are classed as carbon-neutral. Their combustion does not result in an excess of carbon dioxide due to the manner in which they captured the carbon in the first place. This mechanism is called photosynthesis. During photosynthesis, plants take up carbon dioxide from the atmosphere and water from the air and soil, along with energy from sunlight to produce sugars, starches and cellulose, which are the components of the vegetable matter. This process can be represented by an equation:

$$CO_2 + H_2O + \text{Energy (from sunlight)} = [CH_2O] \text{ chemical energy store} + O_2$$

The CH_2O is the carbohydrate molecule, which is a compound of carbon and hydrogen and represents the chemical energy store of the plant material. Carbon dioxide is taken up from the air when the plants are growing and most of it is liberated during the process of combustion. The net result is that there is very little or no excess release of carbon, as carbon dioxide, back into the atmosphere, even taking into account the fossil fuel CO_2, which is released in planting, harvesting, processing and transporting the wood. It is also worth mentioning that wood fuels do not release the acid rain gases (nitrous oxides or sulphur dioxide) when burnt, although carbon dioxide can react with water to form the milder carbonic acid. When new trees are planted (in sustainable forests)

Figure 36: This Farm 2000 straw-burning boiler can burn bales of straw that are almost the full diameter of the boiler. (Author/www.farm2000.co.uk)

to replace the wood that is burned as biofuel, the fuel can be described as sustainable and the lifecycle production of carbon dioxide is close to zero. This is why sustainable biofuels can be described as renewable energy sources.

Table 16: Approximate calorific values of some common fuels.

Fuel	Gross Calorific Value (MJ/kg)
Ethanol	30
LPG	49.4
General purpose coal (5–10% water)	30 to 35
Peat (20% water)	16
Kerosene (28 Redwood seconds, home heating oil)	46.2
Gas/diesel oil (35 Redwood seconds)	45.6
Natural gas	55
Wood (15% water)	12
Wood chips	10.5
Wood pellets	18 to 20
Short rotation coppice	10.6 (as received), 18.6 (dry)
Poultry litter	8.8
Biogas	23
Straw	14

Values shown are approximate only and could vary widely depending on material source and moisture content. Gross calorific value is given in the above table and this value reflects the heating value before the water is driven off. Net calorific value is the heating value of the fuel once the water has been removed.

Source: All figures from *Digest of UK Energy Statistics 2003* table A.1, estimated average gross calorific values of fuels, and 'Log Pile', NEF Renewables, www.logpile.co.uk.

AVAILABLE BIOFUEL RESOURCE

Within Europe, particularly in countries like Austria, Sweden and Germany, there is already a very well-established market and culture for biomass fuels. Austria and Finland, for example, which are heavily forested, use wood in virtually every aspect of life and this underlines the fact that biomass has competing uses: for food, for fodder (straw and grasses) and fibre (for construction, paper, fabrics, etc.). By contrast, the island of Ireland is very lightly forested – with only 6 per cent of the available land area planted as forest in Northern Ireland and 12 per cent in the Republic of Ireland – and most of that is purely for amenity and recreation.

In the UK, the biofuels resource comes from a number of sources and recent studies have indicated that these can make a meaningful contribution to energy requirements in the coming decades. The biofuels resource itself is constituted from waste and forested wood, wood chips, wood pellets, sawdust, straw, a variety of grasses, rape seed, slurry from cows and pigs, waste paper, household waste, landfill gas and sewage, to name but a few.

ENERGY FROM BIOFUELS

A range of techniques can be used to extract the energy from these sources. The simplest is straightforward combustion, as is the case, for example, in wood fires and boilers or in straw burning boilers. The energy in a fuel is given by its calorific value (CV).

Other processes can be used to improve the quality of the fuel and these are termed thermochemical processing. They include pyrolysis, gasification and liquefaction. A further category of energy extraction is biological treatment; this includes processes such as aerobic and anaerobic digestion and fermentation. These processes can be used to produce gas, which can be burned in an internal combustion engine to generate electricity and heat. As can be inferred from the foregoing, the main output from these sources is heat, and unless there is a reasonable scale to the operation, electricity will not be a product. Biofuels are rich in volatiles, perhaps as high as 75 per cent compared with 50 per cent for coal, so special attention should be paid as to how they are burned to ensure that the volatiles contribute to the useful energy before they disappear up the chimney. Unlike most other fuels, biofuels may contain significant quantities of water, which sometimes can reach 95per cent. This means that the useful energy of a biomass fuel may be much less per tonne compared with conventional fuels. Consequently these fuels are usually dried to remove the moisture before burning or transportation. An example is air-dried willow, which contains around 20 per cent moisture, less than freshly cut willow, but more than oven-dried willow.

The potential of several biofuels will be briefly reviewed, with special attention paid to those that currently attract government support through grant aid. Although governments have not announced specific targets in respect of renewable heat, they have supported small-scale heating boilers and wood-burning stoves through grants. With rising fossil fuel prices, fears about fuel security and lack of competition, the sustainable option offered by wood carries a strong appeal, especially to those who are building new properties in the countryside where fuel choice may be limited, but wood and space is more readily available.

Gasification of Wood Fuels

Gasification is the process of producing a gaseous fuel from a solid fuel, which is reacted with steam and air (or oxygen) in a gasifier. There are a number of discrete processes involved and these require a combination of temperatures and pressures to produce the fuel gas. The gas contains methane and hydrogen as the active components, but there are also varying amounts of carbon dioxide, carbon monoxide and nitrogen.

Gasification of solid fuels has been around for a long time – coal-derived town gas was used extensively before the wider-scale introduction of natural gas. Gasification provides the benefits that it is both cleaner, because the pollutants can be removed during its production, and more versatile than the original biofuel. The gas produced burns at a higher temperature leading to higher efficiency. Despite some promising developments in the early part of the twentieth century, the production of electricity from small-scale wood gasifiers has been difficult to achieve commercially, although viable large-scale plants have been built in Europe: Güssing in Austria and Varno in Finland are examples. On a smaller scale, tarry by-products in the flue gas stream have limited the exploitation of this technology using diesel engines. The tarry deposits foul the engines resulting in frequent shut-downs. The use of the biofuel gases in gas turbines, even at a relatively small scale, has attracted some interest, although again the wider deployment of these turbines remains limited.

Pyrolysis

Pyrolysis is the process of heating a biofuel (or waste) at 300°C to 500°C in conditions where there is virtually no air present. This drives out the volatiles from the feedstock leaving the char (or charcoal), which has a higher energy density than the original biofuel. An example of this process is the production of charcoal from wood. In one sense the fuel is improved by pyrolysis, although most of the original energy content is lost with the volatiles unless they are captured. Pyrolysis has been developed to treat a variety of waste streams and can be combined with other technologies to provide an overall comprehensive treatment for municipal and commercial wastes. It could be a commercially viable technology within five to ten years.

ANAEROBIC DIGESTION

Anaerobic digestion, or decomposition in the absence of air, can either take place through the action of bacteria at normal outdoor temperatures, or at a higher temperature. It occurs with almost any organic material in suitably warm, moist and airless conditions. It is an entirely natural process that happens frequently in ponds and lakes, where organic materials decay. The process can be adapted to provide a source of energy, for example biogas, which can be derived from feedstock such as sewage or animal manures (as well as a host of other organic materials and bio-wastes). Landfill gas is a variation of this process and it is produced when wastes are buried at landfill sites. The gas is liberated as the waste decays and, if captured, it can be used as a fuel for a diesel engine and will produce both heat and electricity. The heat is provided by the engine-cooling water jacket and the electricity is produced from a generator coupled to the engine.

With the tightening of legislation restricting the spreading of manures and sewage onto the land, the opportunity to produce useful heat and electricity from a process that effectively disposes of the waste has become very appealing. In anaerobic digesters, the slurry (which may be over 90 per cent water) is collected in a large tank and allowed to digest for a period, dependent on the feedstock material, of anything from ten days to four weeks. During this time biogas will be liberated and collected and, concurrently, there will be two other product streams from the digester. The first is a liquid that may contain nutrients (nitrates and potassium) and the second is a solid, pathogen-free fibre. The liquid can be used as a fertilizer and in some cases the solid may be spread on land as a soil conditioner. The digestion process and the end-use of the by-products are very dependent on the type of waste. Some wastes can be effectively treated and the harmful bacteria and pathogens, together with unwanted odours, will be destroyed so long as the waste resides in the digester for an appropriate period. Biogas liberation usually begins to decline to a steady minimum after twenty-five to thirty days.

Digesters are available in a range of sizes with capacities of a few hundred gallons (for a single farm unit) to multi-farm units of several thousand litres. A well-managed digester will produce around 200 to 400m^3 of biogas, with a methane content of around 50 to 70 per cent.

Digesters can treat unpleasant wastes and at the same time they can aid with the balancing of nitrates and potassium contents in the soils. Experience so far has been mixed, with a number of successful units operating at a range of scales across Europe. On-farm anaerobic digesters have been developed to handle the wastes from individual dairy and pig farms. In the UK and Ireland, take-up has been less enthusiastic, with significant barriers still to be overcome, particularly with regard to finding suitable sites and establishing economic schemes. It is estimated that a digester will require around 20,000 tonnes of waste per annum to make it economic, although it may be viable at lower waste tonnes. Communities of farmers could provide sufficient waste to make digesters economically viable and, at the same time, they could provide an outlet for the digester by-products. A number of demonstration schemes are under construction in the UK and, if they prove successful, then a wider uptake may follow.

Figure 37: The anaerobic digester uses a feedstock of slurry from cattle and pigs to generate methane (and carbon dioxide) gas, which is burned in a diesel engine to provide heat and electricity. (Image courtesy Geoff Hogan, The Biomass Centre, The UK Forestry Commission)

Figure 38: Two-year-old mature willow growing on Rural Generation's Brookehall Estate. The site has been pioneering the growing of energy crops combined with the treatment of sewage close to the Culmore sewage works in County Londonderry. (Author/www.ruralgeneration.com)

LANDFILL GAS

When municipal and commercial wastes are buried deep in landfill, the conditions will be suitable for the anaerobic digestion of the organic fractions. The principal useful gas component is methane, usually in a mixture with carbon dioxide, which offers no energy contribution. Generally the conditions in a landfill site are suitable, but not as amenable as in a digester, so the gas liberation rate is slow. Useful quantities of landfill methane will appear after roughly one year and the site may continue to liberate gas for up to twenty years, depending on its size, depth and waste character. In the past, landfill gas has been flared wastefully, but since the 1970s there has been a move to engineer landfill sites so that the gas can be recovered and used as a fuel. At larger landfill sites, heavy-duty diesel engines can burn the gas to give both heat and electricity, although more recently gas turbines have been used to give better efficiencies.

AGRICULTURAL WASTES

Agricultural wastes, from crops and animals, provide large amounts of energy in many countries across the world. The burning of animal wastes, such as dried dung, is second only to wood burning as an energy source. Agricultural wastes include materials such as mushroom compost, straw, animal wastes (chicken, pig and cattle slurries and food industry by-products), chicken litter and vegetable waste. The vast majority of these wastes can be digested to provide fuel biogas (see above). If the water content of these wastes is low enough, it is also possible to extract the energy by direct combustion in a purpose-built boiler. Demonstration projects have been constructed to burn wastes, such as poultry litter, and some plants already generate several MW of heat and electricity through direct combustion.

ENERGY CROPS

With the economic and environmental pressures on modern agriculture increasing and a reduction in the need to grow food crops, farmers have turned their attention to energy crops. These crops, which are specifically grown and harvested to produce energy, fall roughly into two categories: wood and other plants for burning, and crops for fermenting to ethanol, or whose seeds can be pressed for oil such as hemp or rape seed.

Coppiced willow is an example of a wood energy crop. The principle is that fast-growing trees, such as willow are cut back every couple of years and then allowed to sprout again. This process is referred to as short rotation coppice (SRC). This is an ancient technique that can be applied very effectively. Willow grows well in most northern European soils and it thrives on relatively poor, sometimes undrained land. It is harvested in a two- or three-year rotation cycle, which means that for the first couple of years, until the plant has matured, an alternative fuel will be required. Indigenous willow from the British Isles has relatively good resistance to disease, although the best varieties and consequently those used commercially come from mainland Europe. Planting and harvesting the willow requires specialist equipment, which is usually shared collectively across a number of farms where the willow is grown. The equipment may be hired, or a contract may be set up to ensure it is available, when needed. An energy services company (ESCO) can perform one or more of the activities identified above and could be fundamental in establishing a viable SRC operation.

Harvested willow needs to be dried and chipped before it is burned in a boiler and this will require energy. Recent trials have shown that willows are effective at removing pathogens from sewage cake applied to their roots (known as bioremediation). Although not yet fully proven, if this application proves feasible, it will dramatically improve the economics of willow as a combined energy/waste-treatment crop. Fears still remain about the combustion of this willow due to the presence of heavy metals and other contaminants in sewage wastes, although these tend to be retained in the root systems. Farmers have been keen to embrace willow coppicing, growing the plants and then burning them, to become self-sufficient in their heat requirement. A recent proposal has been the introduction of an ESCO (in the form of a group of farmers to provide large-scale willow harvests) to provide fuel for a commercial or industrial heat requirement. There have been a range of grants available for the planting and for the drying or cutting of biomass.

There is currently considerable interest in the production of biofuel ethanol from a range of plants such as sugar beet and miscanthus (elephant grass). Interest has also been growing among farmers who are keen to diversify into the growing of oil seed crops, which yield vegetable oils. These oils are mostly compounds of hydrocarbon-like fatty acids and glycerol. They can be burned in diesel engines, either in their pure form or after blending with diesel fuel. Energy crops, planted out on 20 per cent of a farm's working land, have the potential to support active farming communities, helping them to live and thrive in the countryside, although the economics of this depends heavily on a range of market influences.

WOOD FUELS

There are a wide variety of wood fuels available and a plethora of appliances suitable for burning them. The options range from logs that can be burned in an open fireplace, to wood pellets that can fuel specially designed boilers to provide hot water and space heating using radiators. This section will focus on wood burning systems that have attracted grant funding in recent years.

Market Potential for Wood Fuels

Some of the most promising areas for the development of wood fuel are:

- Domestic, commercial, community and industrial space heating;
- Industrial process heat;
- Combined heat and power (on a range of scales).

This chapter will concentrate on the first of these opportunities. The scope for log burning is usually limited to rural wooded areas, although urban wood fuel applications, using pellets and wood chips, are gaining importance in space heating. In domestic homes, wood-fuelled stoves and boilers are becoming increasingly common. Fully automated, user-friendly wood chip and pellet boilers are on the increase and the temperate nature of the climate in the UK and Ireland means that pellet-fuelled central heating systems are becoming more widespread.

Figure 39: Wood chips from SRC willow.

Wood fuel technology is now sufficiently developed that larger buildings, such as hotels, leisure centres, hospitals, schools, housing associations and commercial premises, can take advantage of automated wood chip and pellet central heating systems. The lower cost of these fuels (compared with oil, gas and coal) could bring significant fuel cost savings to these particular buildings, especially since they have a high occupancy and year-round heat demand. Storage space and vehicular access to the storage silos may be a drawback in some instances. One way around this, however, is the use of Energy Cabins or plug-and-play modules. These are self-contained pre-assembled mobiles that house the boiler and the fuel storage. They can be towed into position and connected into the existing building's heating system. They may also be combined with other forms of renewable energy, such as solar thermal panels for hot water, or with PV or wind turbines to provide electricity. These wood- and pellet-burning systems normally include a hot water buffer tank, which allows sufficient hot water storage to accommodate any interruption in the combustion process, perhaps due to refuelling or lack of automation. A recent variation of the Energy Cabin has been the liquid biofuelled version, which burns biofuel derived from a variety of sources in a

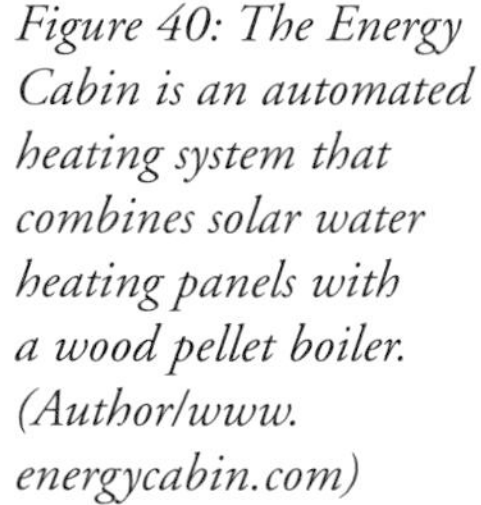

Figure 40: The Energy Cabin is an automated heating system that combines solar water heating panels with a wood pellet boiler. (Author/www.energycabin.com)

combined heat and power engine, to provide space and water heating with electricity.

New residential developments offer great opportunities for wood-fuelled district heating. A central biomass boiler, fuelled with wood chips from purpose-grown willow or even a large pellet boiler system, could provide enough hot water to supply the development through a network of pipes laid down during construction. These boilers can also be either multi-fuel or dual-boiler systems (one biomass boiler and the other oil or gas). This will provide fuel-secure supplies of heat or it will permit economic advantage as the relative price of fuels change. Each property supplied would have a heat exchanger (to supply space heating and hot water), complete with an individual heat meter and a mechanism for billing, depending on the heat consumed. Such schemes could be operated by ESCOs. In this arrangement, a customer might pay for the energy to heat the building and provide the hot water, and for the electricity consumed, as well as for any energy-saving devices that may be fitted. The ESCO might design, construct, operate (to include fuel provision) and maintain the heating system. This could, for instance, be a farming cooperative, who would collectively grow the willow, harvest it, dry it and transport it to the central boiler house.

There are a few fledgling ESCOs emerging and there is no doubt that this managed energy service could provide advantages, including the guaranteed supply of wood fuel during the start-up years when the development would not yet have its own willow fuel established. To date, there have been a number of barriers to the introduction of ESCO-operated district heating schemes. These are largely non-technical, associated with culture and the perception that large central boilers are unreliable and outside the influence of the individuals receiving the heat (many of whom would prefer to own their boiler).

The use of wood fuels to provide process heat and even electricity for industrial applications is less novel. Ballycassidy Sawmills is a prime example of how this can work (*see* page 87). In the UK and Ireland there are a number of companies involved in the wood-processing industry, such as panel board mills, sawmills and joineries, that use wood wastes such as sawdust, pallets, bark and shavings to generate process heat.

LARGE-SCALE BIOMASS DEVELOPMENT

There are several large-scale biomass plants under development that will rely on sustainably harvested forests for their feedstock. These plants use wood to raise steam, which drives a conventional steam turbine to generate electricity. The UK's largest dedicated and fully commercial biomass station, at Steven's Croft, Lockerbie, is owned by the energy company E.ON. The station cost around £90 million and can generate 44MWe of electricity at peak output. In December 2007 the UK government gave permission to construct an even larger 350MWe wood chip-fuelled power plant at Port Talbot in South Wales. This plant will be fuelled by wood from sustainable sources in the USA and Canada. At a cost of £400 million, the biomass plant is due to come on line by 2010 and provide continuous output for 365 days a year. When completed it will be the largest biomass plant of its kind in the world.

BURNING WOOD FUELS

Wood may be obtained from a variety of suppliers and the quantity and type of wood available will be highly variable, sometimes a mix of hard and soft woods. When purchasing it as a renewable fuel, it is essential to determine that the wood is from a sustainable source, whether or not it has been seasoned, the price including delivery, the type and size of the wood and its quantity and volume by weight. It is particularly useful to know whether or not the wood has been reclaimed, since it may have been treated or painted.

TRIMMINGS AND CUT WOOD

Recently felled and cut timber contains around 50 per cent moisture by weight and this gives rise to a number of problems when burned. Poplar is one of the wettest woods when freshly cut; some of the best are ash, beech, hornbeam, hawthorn, crab apple and wild cherry. Although dense wood tends to have a slightly higher calorific value and will burn for longer periods, some dense hardwood varieties, such as oak and elm, will be difficult to burn. It is usual to let wood sit for some time before it is burned and this

Table 17: Wood fuel characteristics.

Fuel Type	**Logs, Coarse and Waste Wood**	**Wood Chips**	**Wood Pellets**	**Bark**	**Sawdust**
Source	Forest cuttings and trimmings, logs, demolition wood, pallets, wooden furniture, doors and windows	Forest maintenance, energy crops (e.g. willow). Fuel store design is key	Wood processing plant waste, pallets, some cereal grain	Sawmills, pulp and paper industry	Sawmills
Moisture content wt % (wet basis)	10–30%	40–60% (after harvesting), 30–40% after storing and before chipping	< 10%	45–65%	40–60% (fresh cut), 10–20% (dry)
Energy density (kWh/m³)	Low	600 (High)	3,450 (High)	Low	Medium
Key characteristics	Easy to handle, can be produced from small-scale woodland.	Automatic feed for 24-hour operation. Large-scale operation most economic. Need local supplier	Transportation over distance straightforward. Fuel delivery by blower	High in ash and irregular quality	Transportation over distance straightforward
Size and shape	Variable	20–80mm	6–8mm in diameter and 5–30mm long (say 2.5cm)	Extremely irregular shapes and sizes	< 5mm
Impurities	Treatment chemicals, nails and paint	Bark contamination. Sometimes mineral contamination	Maize starch binder in small quantities		Sometimes sandpaper or grindstone residue
Ash content wt % (wet basis)	1.5–12%	0.5% (softwood) to 2.5% (hardwood)	< 0.5%	5–8% (delivered) 2–5% (uncon-taminated)	0.4–1.1%
Quality standard	Under development by CEN TC 335 and CEN TC 343	Under development by CEN TC 335	Under development by CEN TC 335	Under development by CEN TC 335	Under development by CEN TC 335

Source: Based on data extracted from Log Pile, NEF Renewables, www.logpile.co.uk.

is referred to as seasoning. Depending on the type of wood and the time of year when it is cut, it may be better to let the wood season for a year or more. Freshly cut wood is best stored under cover in an airy open-sided housing. When the moisture content approaches air-dry (around 25 per cent moisture or less) the wood is ready to burn. Some of the heat from combustion is used to turn the water in the wood into water vapour and this heat is lost in the flue gas. This will increase the volume of gas in the combustion chamber and flue, and burning wet wood must allow for this. Wood is ready to burn when the bark peels away easily and is noticeably dry. Combustion-ready logs will have cracks across the grain and they should be no more than 10 to 15cm thick.

LOG BURNING

In log- and wood-burning appliances combustion takes place in a single chamber (the combustion chamber), which contains the fuel and the primary and secondary air. The fuel is introduced manually, in batches of wood and logs, or as single logs. In through-burning boilers and stoves, combustion air enters the furnace from below the logs and they are ignited from the bottom to ensure good burning of the whole stack. In top-burning log appliances there is no grate and the combustion air enters from the side of the furnace. The process is less effective than through-burning and in both cases the combustion air is introduced by natural draught (that is, there are no fans to blow the air in). A recent development in these appliances is the use of under-burning, where the stack is ignited at the bottom and the combustion gases are drawn downwards or sidewards. The logs burn one after another giving a relatively continuous burning process. Under-burning appliances require a fan on the furnace exit to suck the combustion gases out and push them up the flue, which means that the furnace operates under negative-pressure.

The use of wood as a boiler fuel has increased with the advent of automatic control for the fuel feed and boiler control systems. Logs have been burned in wood-burning stoves for some time, but new designs allow them to be used in boiler systems feeding radiators in a central heating system. Log stoves have efficiencies around 65 per cent, but modern log-fired boilers can achieve 90 per cent efficiency. Log boilers need to be filled with logs at least once a day and require frequent ash removal. Log boiler systems that are used to heat radiators usually include a hot water buffer storage tank.

WOOD CHIPS

Wood chips can be processed from sustainable forests and purpose-grown energy crops. They are irregular in shape, although some chipping machines do produce reasonably uniform chips. Consistency in wood chips is essential since a sudden change in wood quality, size or moisture content will affect the boiler output and could completely extinguish it. Chips will have varying water contents and are frequently dried before use. They are much less dense than wood pellets and accordingly do not lend themselves so well to automated control and feed systems, although these are available.

Wood chip boiler systems start around 25kW, which means that they are well suited for larger heating loads (commercial premises, schools, community applications) and they are ideal for district heating schemes. As a fuel, their strength lies in two areas: low cost and availability in rural areas. In the countryside, storage space is generally more readily available and disposal of ash less of an issue, so they are ideally suited as an environmentally friendly fuel option, especially for the rural fuel poor (those who have difficulty paying their energy bills). Wood chips are usually stored in a silo or dedicated storage building close to the boiler and they are fed to the burner by an auto-feed mechanism. It is more sustainable to have a local network of producers to ensure that the wood chips are not transported long distances, which is particularly important if frequent refuelling is necessary.

WOOD PELLETS

Wood pellets are manufactured using the by-products or wastes from the sawmill industry and other wood processing plants, such as scrap pallet processing. Their properties make them an ideal fuel for heating systems of all sizes – from small domestic room heaters or stoves, to larger industrial boilers. Compared with

Figure 41: The Ballycassidy Sawmills (Balcas) 'Brites' wood pellet facility at Enniskillen in County Fermanagh can produce around 50,000 tonnes of pellets per annum. It uses waste wood to fuel the CHP plant. (Photograph courtesy of Balcas)

other coarser wood products, such as logs and chips, pellets have a lower volume (on an equivalent energy basis) for transportation and distribution. Wood pellets require less storage volume (due to their high energy density), need fewer deliveries and can be used in a range of heating appliances. They are low in ash and relatively clean from an emissions viewpoint, and can generally be stored without degrading. Pellets are easy to ignite and should have a good consistency of size and moisture content. They are easy to handle and may be fed automatically from storage hoppers, and can be auger- or screw-fed to boilers automatically. As a renewable fuel, wood pellets are exempt from the UK Climate Change Levy.

Wood pellets have been available for some time in North America and parts of Europe, but they are a relatively new fuel to the UK and Ireland. Since 1999 a number of manufacturing facilities have been constructed, including Welsh Biofuels in South Wales, Premier Waste in Durham and, on a smaller-scale, Renewable Heat and Power Ltd in Devon. In 2005 the construction of a major pellet factory at Ballycassidy Sawmills (Balcas) in Enniskillen, County Fermanagh, made bulk wood pellets available from a local manufacturer for the first time in Ireland. The Balcas project uses wood waste from the main sawmill plant to fuel a combined heat and power (CHP) system to produce 2.7MW of electricity, 10MW of process heat and 50,000 tonnes of wood pellets. There are plans to open a number of similar, new pellet plants in the UK, and ClearPower is scheduled to complete a major pellet manufacturing facility in Ireland during 2008.

Pellets are cylindrical with a diameter between 6mm and 12mm and a length of around 30mm. Those over 25mm long are sometimes referred to as pellet briquettes and are used almost exclusively in industrial markets. Most pellets are made from good-quality, untreated wood waste or sawdust and they are held together purely using the lignin naturally provided by the wood (although in certain instances it may have to be supplemented). They burn cleanly, although not smokelessly, and this has implications for where the pellet heaters and boilers may be used. They are low in moisture, usually less than 10 per cent, and are compacted so their energy content is

Figure 42: Pellets are available in 10kg or 25kg bags that are suitable for domestic boilers or pellet stoves. (Photograph courtesy of Balcas)

high. Pelletizing improves the form, density, transportation characteristics and handling capability of the fuel but does not improve the calorific value. The amount of pre-drying during manufacture will determine how much energy is used, or lost, to the process. With some timber by-products, a moisture content of up to 30 per cent or above is not unusual and this will require energy to produce a drier product.

Good quality domestic pellets are manufactured from forestry and agricultural waste products that have not been treated or contaminated by any waste or manufacturing process. Commercial grade pellets are derived from similar feedstock, but may also include vegetable waste from agriculture and forestry, and perhaps the food processing industry. Most forms of wood waste are suitable, including cork, so long as it does not contain halogenated organic compounds or heavy metals as a result of treatment

Wood Pellets

Wood pellets have both advantages and disadvantages when compared with other wood fuels.

Advantages

- Pellets are homogenous and consistent in moisture content and structure.
- Pellets are easy to ignite.
- Pellets are compressed, therefore they have a lower volume (than bulk wood such as logs or wood chips) to transport and store.
- Pellets have a high energy density.
- With appropriate storage, pellets require fewer deliveries.
- Pellets can be used in stoves and boilers.
- Pellets flow like a liquid and can be used with automatic feeders and controllers.
- Pellets are easy to handle.

Disadvantages

- They must be kept dry – moisture will lead to pellet disintegration.
- They can have harmful contamination – additives, dirt and other farm wastes.
- Pellet consistency must be maintained to ensure steady running of the boiler without shutdowns.
- Rough handling of pellets can lead to excessive dust (fines). Too much dust can block the auger feeder and choke the boiler.
- They will be more expensive than other forms of less refined wood fuels.
- There are still a limited number of fuel suppliers, although this situation is improving.
- Bulk deliveries are very often in set amounts.

(or painting). These contaminants usually come with timber reclaimed from the building industry or from demolition waste. Pure wood should be used that is free from contaminants such as plywood or chipboard glues. Pellets should not contain additives apart from lignin binder or, occasionally, trace amounts of vegetable oil, which is used as a lubricant in the die presses. Small quantities of maize starch are sometimes used during the manufacture as an additional binding agent. Pellets do produce a small amount of ash and they can be categorized as Premium (with up to 1 per cent ash), Standard (with 1 to 3 per cent ash) or High ash (with 3 to 6 per cent ash).

As noted, burning wood pellets produces smoke, so they are not classed as smokeless fuels. The relevant UK legislation is detailed in the Clean Air Act, which 'prohibits the emission of dark smoke from chimneys, including those associated with domestic premises'. The Act also specifies chimney heights and allows local authorities to declare smoke control areas (or smokeless zones). In these areas all combustion appliances (boilers and heaters) must be approved as smokeless, or they must burn smokeless fuels. Only a few woodchip and pellet boilers have this approval (these are designated as exempt appliances), although new equipment can be tested and approved.

Pellet Quality

Experience to date has shown that pellet quality is probably the single most important issue for satisfactory operation with a pellet boiler system. Quality can vary widely and as a general rule the cheaper the pellets, the lower the quality. Boilers are designed to burn pellets of a specified quality and any variance will lead to problems. The quality standards are well established in European countries with Germany, Austria and Sweden taking the lead. There is an anticipated European quality standard for pellets in draft, but as yet it has not been forthcoming and various individual countries have their own standard. In the UK the DTI adopted the Code of Good Practice for Pellets (COGP). This code designates two grades of pellets, domestic and commercial, and stipulates raw materials as well as chemical and physical properties.

Pellet Transportation and Storage

A successful pellet distribution network is necessary for the widespread uptake of this product. Pellets are available in small bags for ease of handling for domestic appliances such as room heaters and stoves. Bags of 10kg or 25kg are supplied on pallets, or they can be supplied singly to manually feed stoves and mini hoppers, which hold around 300kg. Pellets

Figure 43: Domestic quality wood pellets shown in the local storage hopper of an Irish-made Woodpecker pellet boiler.

Pellet Quality

The following characteristics are useful as an indication of pellet quality:

Size: pellets are around 6–8mm in diameter and roughly three to four times this in length (say 2.5cm long).

Moisture content: This should be 6 to 10 per cent.

Energy content: 4.5kWh/kg, or around 18 to 20MJ/kg.

Ash content: Should be below 0.7 per cent ash.

Mechanical durability: This is an indication of how well pellets can stand handling during transportation and delivery to site. Pellets will break up and revert to dust when they are moved or impacted, which will increase the amount of dust (or fines) in the fuel.

Pellet dust: Blowing pellets from bulk deliveries can result in an increased amount of fines in the storage silo. Blowing pellets at too high a pressure during delivery can result in the disintegration of otherwise good quality pellets. If there is excess dust in the fuel, it will choke the auger and this will result in stoppages in the fuel supply to the boiler. The dust will also change the heat flux in the fire box (or combustion chamber), which can result in overheating of the boiler materials and ash sintering. Sintered ash is a lump of material that will form on the boiler metal surfaces, restricting heat transfer, creating hot spots and possibly interfering with pellet transportation. The amount of fines should be below 1 per cent, and although pellets are screened to remove dust before they leave the production facility, some users have been using additional screening before the pellets enter their local pellet storage. Bagged pellets should have much less dust than bulk-delivered pellets.

Additives and binding: Pellets should have an absolute minimum of additive chemical material. They are manufactured from homogenous sawdust and their natural lignin is used to bind them. If additional binder is necessary it should be declared. Pellets should typically have less than 300ppm sulphur content and less than 800ppm chlorine content.

Bulk density: This is a measure of the weight of a certain volume of loose wood pellets and should be in the order of 650kg/m^3. If the weight is too low then the pellets have not been pressed hard enough during manufacture and they may dissociate easily to dust.

Inclusions: Sometimes visual inspection of pellets can reveal the presence of unwanted material, such as darker or coloured particles. Some pellet manufacturers include cereal grain and mixed qualities of wood or bark, which can lead to boiler shutdown or increased ash and smoke. Pellets may be dark brown on the outside, which is a sign of excessive friction in the presses during the extrusion process, but so long as the pellets are not brown on the inside this is not a serious problem.

Water Testing 1: A sample of pellets may be placed in a container of water to test for excessive or incorrect binding. If the pellets do not rapidly dissolve into sawdust within minutes, then they are likely to contain too much binder or the quality of the wood feedstock is poor. Once the pellets dissolve in the water, the heavier components will fall to the bottom of the container. If there is an excessive amount of heavier dust, which may be a waste product from a sanding operation (resulting in sandpaper waste), then there may be more ash and a higher probability of sintering in the boiler.

Water Testing 2: A container of around 2 litres should be weighed on a kitchen scale, then filled to the brim with pellets and re-weighed and both weights noted. The empty container should now be filled with water and the weight noted. The weight of the container should be deducted from both of the previously noted measurements. The weight of the pellets should be divided by the weight of the water to give a figure of around 0.6 to 0.7. If this ratio is below 0.6, then it is an indication that the pellets have not been properly pressed and low durability will result with high levels of dust likely.

(Extracted from Pieter D. Kofman, 'Simple Ways to Check Wood Pellet Quality', *Bioenergy News* [SEI REIO] (Winter 2006–7), pp. 8–9)

Figure 44: Three tonnes of pellets are air-blown into the external storage tank from a bulk delivery tanker. (Photograph courtesy of Balcas)

purchased in this way are significantly more expensive than when purchased in bulk. A few manufacturers offer larger bags and these need to be moved and stacked by small cranes or forklift trucks. The larger bags are mostly for retailer distribution or for farm use.

Bulk Pellets

Bulk loads are delivered by special tankers, in the same manner as fuel oil, and the pellets may be air-blown into storage silos. Pellets require about three times the storage space of oil for a similar heating demand, although this is only an approximate relationship. They are normally housed in a pellet store within 5 metres of the boiler. Storage of pellets should be in specially constructed, completely waterproof storehouses, or they can be held in metal silos or plastic tanks. Sometimes steel frames supporting an anti-static material can be used as an indoor pellet silo. Stores should hold around four tonnes of pellets, as this will allow a standard bulk delivery without running out of fuel. Storage tanks and silos are usually constructed with sloping floors to facilitate pellet delivery to the fuel draw-off conveyor. Screw conveyors (or augers) are used to deliver pellets to the boiler or alternatively they may be fed using a suction system.

WOOD FUEL ECONOMICS

The capital costs of installing a wood fuel central heating system are slightly higher compared with the installation of oil and gas systems. However, the fuel costs are lower and relatively stable over longer periods and this should lead to reasonable payback periods to offset the initial capital cost. Although they are independent of international energy prices, wood fuel costs can be variable, depending largely on the accessibility of the fuel supply. In rural areas trimmings and forest waste may be more available, whereas in towns, bagged logs and sticks may be purchased from fuel suppliers (including petrol stations). Logs burned directly in a boiler or stove may cost nothing, or they may have to be purchased in bags. Typical costs might be £45 to £50 for a 300kg bag (logs with around 30 per cent moisture). If they are burned in a wood-burning stove at 70 per cent efficiency, then they will cost around 5p/kWh. Rough wood bought at £30 per tonne with 30 per cent moisture content might cost around 1p/kWh to 2.5p/kWh.

Wood chips with about 30 per cent moisture are available from suppliers within the UK and Ireland at a cost of around £45 per tonne (at 2008 prices). For bulk delivery, with higher moisture content, this

Table 18: Typical costs of a pellet boiler.

Boiler Size (Rating in kW)	Approximate Average Cost (£ 2008)
5kW	£1,900–2,800
11kW	£3,000–3,900
25kW	£3,900–5,500

price might be lower and may be strongly influenced by the closeness to the supply. The energy price of wood chips might be between 1.5p/kWh and 3p/kWh.

Pellets can be manufactured in a relatively small facility, for example a 300kg/h pellet mill running for twenty-four hours per day can produce around 2,000 tonnes per annum. Such a machine costs about £80,000 and there are a number of these small-scale facilities now operating in response to the increase in the number of pellet-burning stoves and boilers being installed. Pellets are available (in 2008) for around £115/tonne to £120/tonne bulk delivery, or £150/tonne to £175/tonne bagged. Assuming a boiler efficiency of 85 per cent, this gives the delivered price for energy of 3.0p/kWh to 4.5p/kWh at 2008 prices. This would put wood pellets on a cost equivalent with natural gas (on a delivered energy basis) and roughly 15 to 30 per cent saving against heating oil, although all these figures are subject to significant seasonal and other more unpredictable variations. (For information on the availability and current price of the various fuels, see the website www.logpile.co.uk.)

Wood Burning Boiler and Wood Stove Grants

Capital grants have been available in the UK and Ireland for the installation of wood-burning boilers and stoves. They are frequently subject to change and have been paid for the installation of approved products using accredited installers; this has ensured a reasonable level of quality. Grants have not been paid retrospectively, so it is essential to secure a grant offer before commencing installation. (For details of grants websites, *see* pages 25–27 and 104–105.)

WOOD COMBUSTION

Stove and Boiler Design

Central heating and general-purpose boilers can be classified in two basic groups – cast sectional boilers or welded sheet metal boilers – and subgrouped as negative pressure and positive pressure boilers. Negative pressure boilers can be naturally ventilated or have fans for forced ventilation.

Welded sheet boilers are assembled as a complete unit in the factory and tend to be less expensive than a sectional boiler. They are normally the most popular choice, but in instances where access is difficult, sectional boilers may be used. Sectional boilers are made of cast iron and are assembled from pre-cast sections on site. On some models it is possible, although potentially expensive, to enlarge the heat transfer area of the boiler by adding more sections to the boiler. This could be necessary, for example, when fitting a pellet burner to a relatively new coal boiler.

In negative pressure boilers the firebox is kept under a negative pressure because the combustion air is sucked through by the natural chimney draught or drawn out by a fan. Pellet boilers generally use naturally ventilated boilers. An ample supply of air for combustion must always be available for combustion appliances, so boiler houses or rooms that contain wood-burning boilers should have good ventilation. The boiler is designed to offer as little resistance to airflow as possible and the flue is used to provide the chimney draught. This means that the correct design and height of flue is essential. Experience has shown that this is one of the most important areas for consideration when designing a wood fuel boiler installation. If there is insufficient draught produced by the flue, then a fan will be needed between the boiler and the chimney but this will increase the electrical losses associated with the boiler as well as increasing the initial cost. Some designs of fan control include an inverter, which can minimize the losses associated with the fan. The flue may have a cyclone flue gas clean-up device to remove the particulates and reduce the flue gas speed. Positive pressure boilers tend to be used in oil and gas applications and are not in common use with wood fuels.

Combustion of the wood fuel in the boiler is controlled by the air flaps, the air fan (if fitted) and the speed of the fuel feeder screw (or auger). Indications of combustion quality will come from: the temperature readings along the gas pass (from the combustion chamber, through to the flue gas exit at the chimney); the excess air ratio (lambda); and the carbon monoxide content (CO sensor) of the flue gas. A boiler electronic controller is used to optimize combustion, using measurements of these parameters. Boiler fans can either be on the combustion air inlet side (positive pressure or over-pressure furnace), or on the flue gas outlet side (negative pressure or under-pressure furnace). Combustion air is introduced to the furnace in two stages, primary air is injected with the fuel at the burner and secondary air is introduced into the combustion chamber to ensure burnout.

BULK WOOD-BURNING STOVES

Wood-burning stoves are becoming increasingly popular (helped by grant support) and they are more efficient than open fires, which lose a lot of heat straight up the chimney. Modern wood-burning stoves have built in secondary air and tertiary air combustion systems, which considerably reduce the unburned carbon. This minimizes smoke and tarry deposit formation, which frequently lead to problems for a range of wood-burning appliances.

Figure 45: An 8kW Rika pellet stove. These stoves are used as room heaters and burn bagged pellets to produce very little ash. They require a flue and should be mounted on a fire-proof base. (Author/www.rika.at)

WOOD FUEL BOILERS

There are three main types of biomass boiler: wood pellet boilers, wood chip boilers and log boilers. When installing any of these appliances it is critical that their storage requirements and the heat demand (sizing) are properly assessed. There are many suppliers of wood-fuelled boilers available, mostly from Scandinavia, Austria and Germany, but there are now a few manufacturers in the UK and Ireland. Biomass boilers are suitable both for new buildings and for retrofitting into existing buildings, and they can be integrated with existing space heating distribution systems (radiators or under-floor heating). They operate in the same way as fossil fuel boilers and can directly replace an oil, coal or gas boiler. Some conventionally fuelled boilers can be converted, by fitting a new burner (and, if necessary, additional heating sections as previously described), for use with wood pellets. For pellets the storage space required is approximately three times the storage space of oil and for wood chips it is approximately thirteen times the storage space. Most biomass boilers are installed in rural locations.

Correct boiler sizing is crucial to ensure that the optimum efficiency is achieved in operation. System performance can be enhanced using a buffer tank and

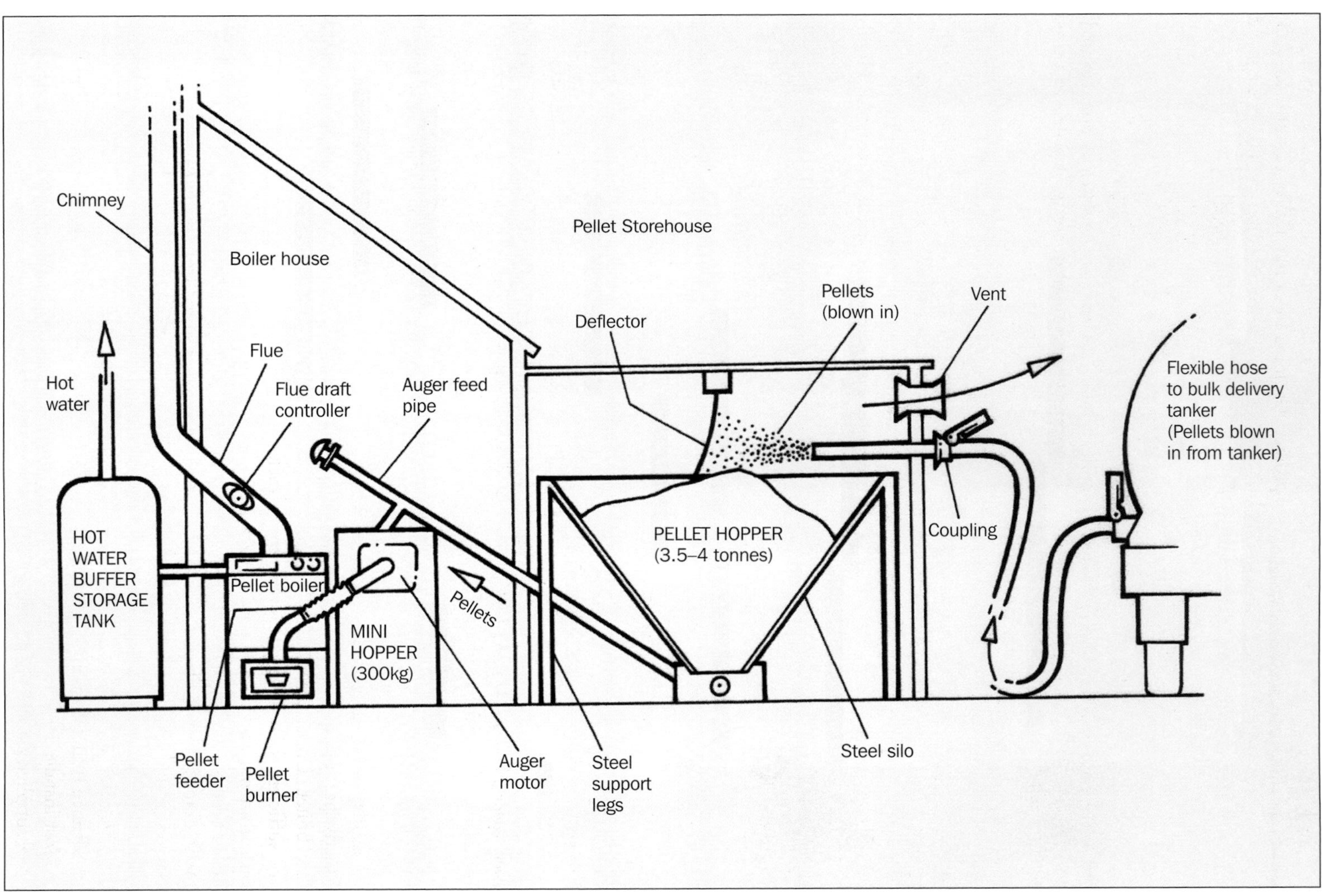

Figure 46: Wood pellet boiler and store.

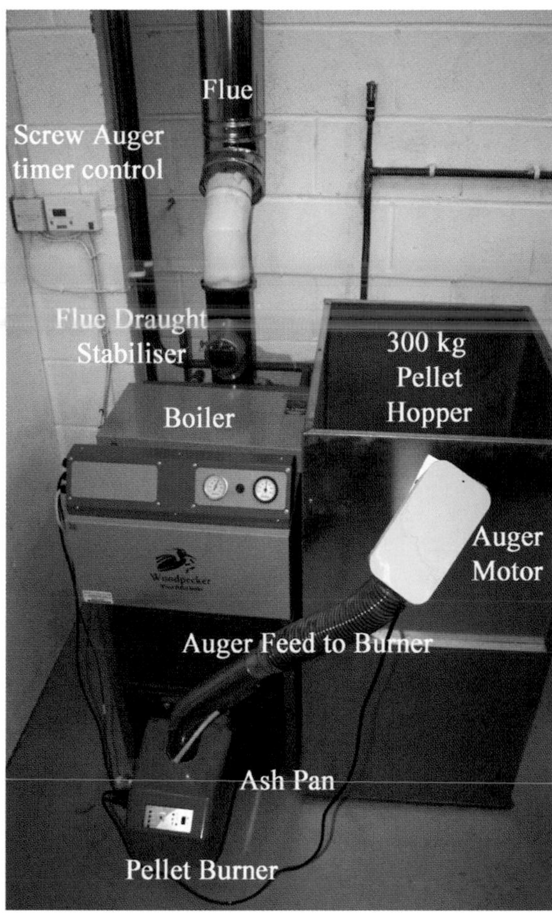

Figure 47: A Woodpecker 30kW wood pellet boiler, showing the chimney with flue stabilizer and mini hopper pellet store. (www.gerkros.ie)

ensuring that generous building insulation is in place. It is very important that the heating circuit pipework is also sized correctly, particularly the manifold pipework around the boiler. Undersized pipes will not be able to carry sufficient heat from the boiler to meet the demand from the building. A product list of the wood-fuelled boilers that have been approved, under government grant schemes in the UK and Ireland, is available on several of the websites listed on pages 25–27 and 101. Those wishing to install a wood pellet, wood chip or log boiler should contact possible fuel suppliers in advance to determine their best fuel supply options. Most installers will be able to recommend a reliable fuel supplier. Boilers and stoves should be mounted on tiled floors, glass or metal plates to protect the floors from the heat of combustion. This is particularly important indoors where boilers and stoves might be situated on wooden floors.

WOOD PELLET STOVES

Wood pellet stoves operate in a similar way to pellet boilers; bulk fuel storage, however, is not required. The stoves burn a small quantity of pellets at a time and use a fan to circulate the heat around the room. They are the most convenient wood-fuelled system to use in a domestic application, although they do require a properly installed flue and adequate room ventilation. Fuel supply is not an issue, since there are numerous suppliers who sell bagged pellets and therefore less fuel storage space is required. Highly efficient pellet stoves are generally used to heat one room and they require little maintenance. Their ash pans need to be emptied roughly every three to four weeks.

WOOD CHIP AND PELLET BOILERS

This section looks at boiler design and considers some of the important components in a little more detail. Automatic fuel feed requires a specially designed burner and these are available for wood chip and wood pellet appliances. The burner contains a number of components, including a feed unit, the combustion retort or grate, an ignition system, a fan, the combustion air system, a controller and several safety devices. The controller manages the burner start-up, ignition, fuel feed and air flow based on the demand for heat. Pellet burners are divided into three main types dependent on the feed direction of the fuel:

- Top feed (fuel fed from above);
- Underfeed (fuel fed from below);
- Horizontal feed.

The fuel is automatically fed by a screw auger delivery system from a hopper to the burner. The ash pan needs emptying every 5 to 6 weeks (or less frequently with some boilers).

Figure 48: A wood pellet boiler overfeed burner.

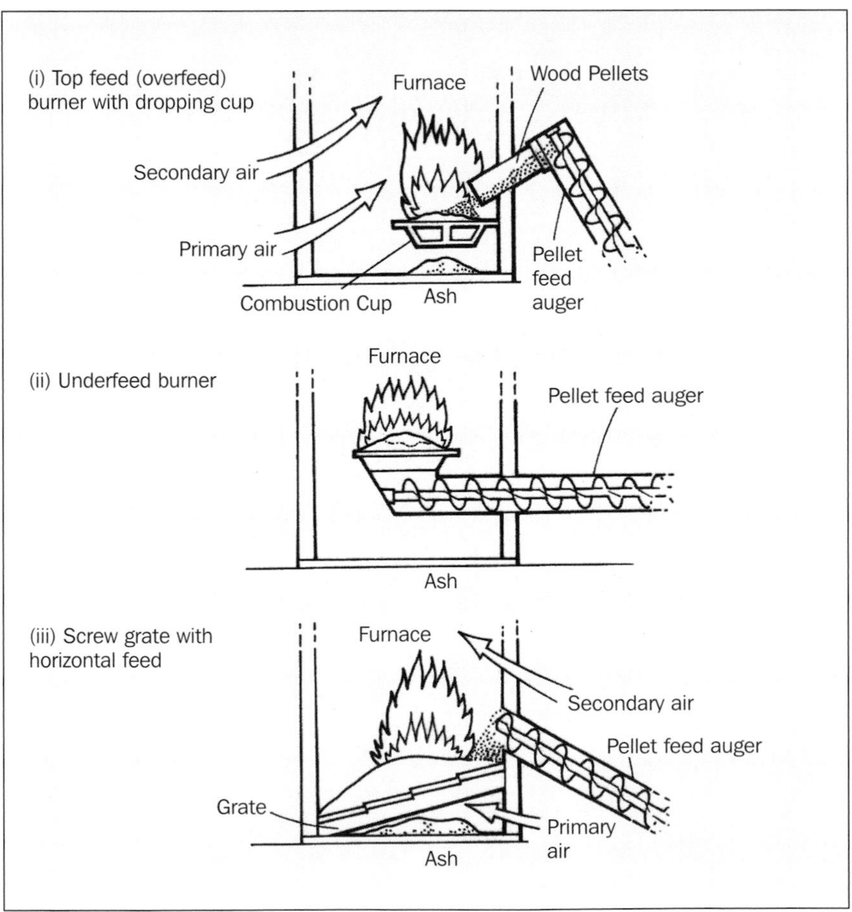

Figure 49: Wood pellets can be overfed, underfed or horizontally fed.

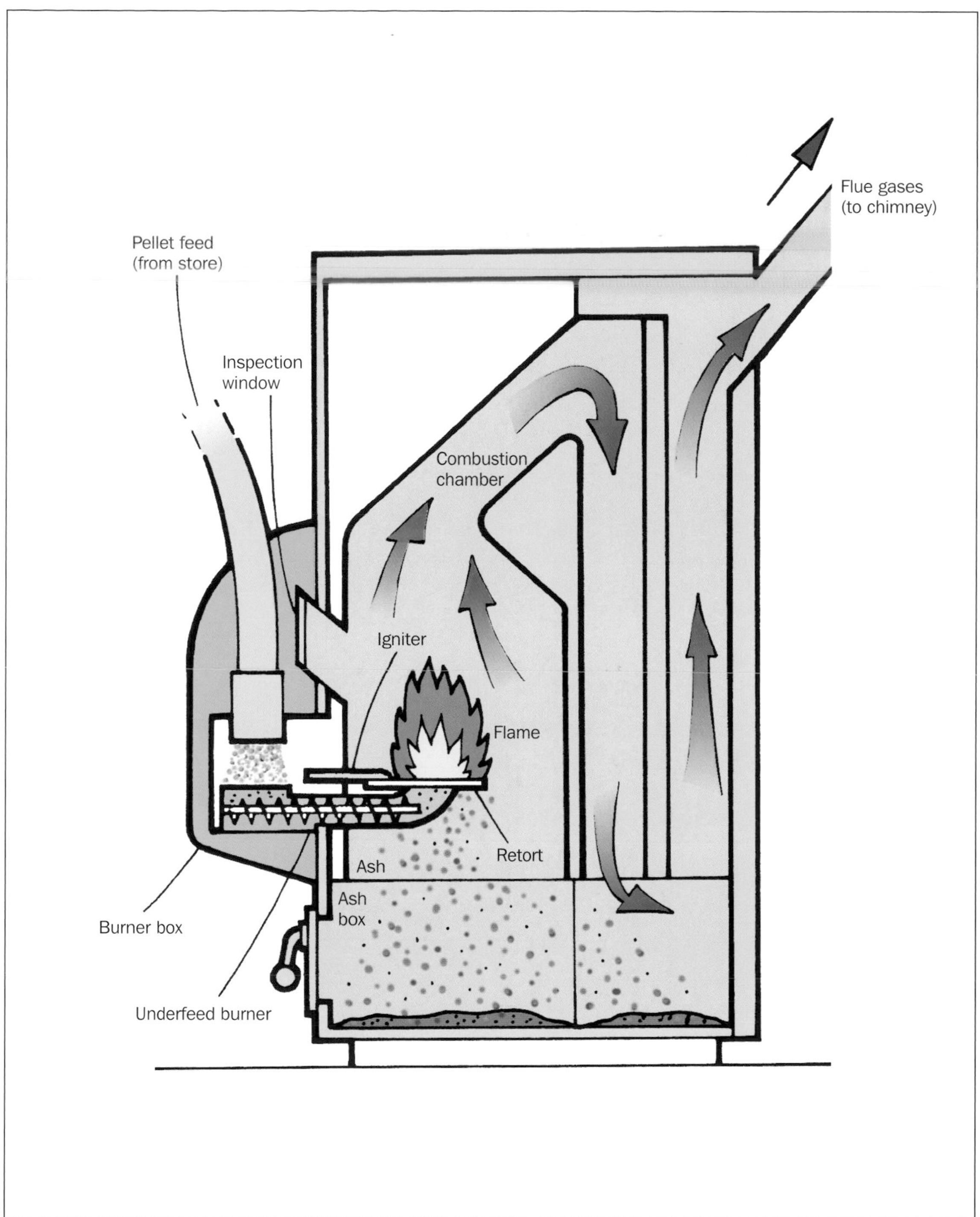

Figure 50: In an underfed system wood pellets are introduced to the retort from underneath and ash spills over the edge into the ash pan.

Table 19: Pellet burner feed systems.				
Burner Type	**Application**	**Principle**	**Combustion**	**Ash**
Top feed	Fitted in small-scale units that burn wood pellets and high-quality wood chips on a retort. Gives accurate feeding for low heat flows of 6–30kW	Pellets are top fed and drop onto the fire bed. They sink to the bottom of the bed. Combustion air rises through the bed	Fuel has long residence time with good burnout. The boiler may not have a grate	Removed manually or mechanically
Under feed	Suits compact wood chip and pellet appliances. Suitable for 10kW to 2.5MW	Pellets are fed to the bottom of the combustion retort. They move up through the fire bed and the ash falls over the edge of the retort into the ash pan	Primary air is introduced from the furnace bottom with the fuel. Secondary air comes into the furnace through an air ring above the burner	Produces small amounts of ash that can be removed manually or mechanically from the ash pan
Horizontal feed	From 15kW to 20MW with a range of options for grates and water-cooled furnaces	Screw feeder carries fuel into the cast iron burner, which may be refractory lined or water-cooled. Burner temperature may reach 1,000°C	Wood pellets, wood chips, sawdust, grain, corn, peat and straw can all be burned with this burner type	Removed manually or mechanically

Table 20: Typical 20kW and 50kW wood pellet boiler specifications.		
Boiler Parameter / Rating	**20kW**	**50kW**
Dimensions (mm)	1,150 × 1,150 × 750	1,170 × 1,580 × 750
Efficiency (%)	85	86
Burner	PX20	PX50
Boiler material	Mild steel	Mild steel
Water content	65 litres	75 litres
Weight (dry)	205kg	245kg
Footprint (mm)	1,150 × 790	1,150 × 790
Tapping sockets	25.4mm	38.1mm

Retrofitting a Pellet Burner

It may be possible to fit a pellet burner to an existing boiler, but it is important to check it for suitability. This would include the boiler condition, its expected output, anticipated flue gas volumes and temperatures, the size of the firebox and ash pit, and the available space around the boiler. Pellet boilers require a large firebox to allow the pellets sufficient residence time in the firebox to burn out. They also require sufficient heat transfer surface to carry the heat to the water. A hydrostatic pressure test and close inspection is recommended before any retrofit is attempted.

FLUE GAS SYSTEM

One of the most important aspects of boiler design is the flue gas system. The products of combustion must be removed from the boiler through a properly constructed chimney and an adequately sized flue. The flue gas system is designed to remove the flue gases and expel them to the atmosphere at a sufficient height to satisfy the environmental legislation. The chimney is the structure that encloses a channel for the flue gas flow. The hot gases leave the boiler and rise by natural buoyancy up the flue. In some boilers a gas flow is maintained through a fan, although the energy it consumes is a loss to the system efficiency. Insulated chimney liners are required for wood-burning stoves and boilers because the temperature of the combustion gases is lower than it would be for an open coal fire. The liner will help maintain the temperature of the gases so that they will rise up the chimney. Anything that cools the gases, such as air leaks or poor combustion, will reduce the effectiveness of the chimney since the buoyancy of the gases will be lower. Flue liners are usually made from fire clay, refractory-quality concrete or corrosion-resistant metal, such as high-grade stainless steel.

Pellet boilers are more environmentally friendly than fossil-fuelled equivalents, but they are still required to provide acceptable levels of flue gas pollutants, including smoke. The system may well include a flue gas fan to maintain a negative draught, a boiler damper for improved efficiency and flue dampers, which are used to keep the negative pressure in the boiler constant. Dust collection devices

Boiler Power Ratings

The boiler power output is dependent on which fuel is used. The power output units can be expressed as kW (kilowatts), btu/h (British Thermal Units per hour) or Mcal/h (megacalories per hour).

Table 21: Approximate conversions of boiler ratings in the various units.

kW	btu/h	Mcal/h
5kW	17,000btu/h	4.3Mcal/h
12kW	40,000btu/h	10Mcal/h
30kW	100,000btu/h	26Mcal/h
50kW	167,000btu/h	43Mcal/h

Older boilers may not have a rating plate that states their output, although this can be estimated from the area of the heating surface. The rating for an oil boiler fitted with a pellet burner will be less than the original rating on oil. An important parameter for the boiler design is the flue gas exit temperature (or outlet gas temperature). Ideally this should be as low as possible to improve efficiency, but if it becomes excessively low, then acids or tarry deposits will condense from the flue gases. As a general rule, the flue gas temperature falls by 5°C to 10°C for each metre of flue length. This means that to prevent condensation in the flue, the temperature should be at least 80°C one metre below the flue exit; indeed, some specially designed flues (with high temperature fans and dampers) can stand much higher temperatures. Boiler exit temperatures are normally around 350°C and it is important not to exceed this temperature.

(Extracted from *Wood Pellet Boilers: A Guide for Installers*, produced by PROWE, a European project under the ALTENER programme, available from the Powys Energy Agency, 2003.)

may also be included, such as dynamic separators, cyclones, scrubbers, electronic filters or barrier filters.

As noted above, it is crucial that a chimney of sufficient height is used. There are various factors that determine what the minimum height must be, but boilers up to 50kW are controlled by Part J of the UK Building Regulations (Part L in Northern Ireland). The basic regulation is that the flue should extend at least 600mm above the roof ridge, but this is conditional on circumstances such as the roof type (flat or pitched) and proximity to adjoining buildings and opening windows. Larger pellet boilers (in excess of 50kW) will be subject to additional legislative requirements. Once the flue has passed through the exterior wall, it is important that there are no horizontal runs (longer than 150mm) and that any bends fitted to the flue are at 45 degrees (not right-angled).

EMISSIONS FROM BIOMASS APPLIANCES

Emissions from wood fuel combustion are similar to those from fossil fuels, without the harmful acid rain oxides of sulphur (SO_x) and nitrogen (NO_x). The principal gases in the flue are carbon dioxide and water vapour, with varying amounts of nitrogen and oxygen from the combustion air. The gases also contain volatiles, trace amounts of carbon monoxide and unburned carbon as particulates. Incomplete combustion will produce smoke and fly ash, unburned particles in the flue gas emissions; this can be due to a number of problems including poor installation, poor quality fuel or a faulty burner. These deficiencies can also result in unburned particles in the ash, which means the ash pan will fill rapidly and heavy, tar-like deposits will accumulate in the chimney. An excess of these tarry deposits can result in dangerous chimney fires. Smoke is harmful to humans, particularly the very small particulates in smoke known as PM 10s, which are linked to cardiovascular illnesses. Lots of smoke, especially black smoke, indicates that the system is not functioning properly. If the smoke billows back into the room (known as spillage), then there may be insufficient draught or a blockage in the chimney and this is a dangerous situation that requires immediate attention. If the chimney produces insufficient draught, a fan may be necessary to draw the combustion gases out of the boiler and eject them to the atmosphere. Carbon monoxide is a poisonous gas that can be produced in systems that are incorrectly installed or operated. Boilers and stoves should always be fitted professionally by an accredited installer and be annually maintained and inspected (including sweeping the chimney). Burning wood produces only small amounts of ash, typically less than 1 per cent, and this can be spread on the garden as a fertilizer.

Biomass boilers can only be installed in smoke control zones if they are exempted appliances. In the UK and Ireland there are a number of exempted wood-burning stoves available from a range of companies (as of 2008) including Clearview, Vermont Castings, Dovre Castings and Dunsley Yorkshire Stoves, and some wood-burning boilers, such as those supplied by Binder, Mawera (UK) and Talbott's C range of boilers.

BOILER SIZING

Traditionally the whole area of boiler sizing has been subjective and rules of thumb are applied liberally, as suggested below (*see also* Chapter 1). Typically a value of $50W/m^2$ is assumed for a domestic house. This has led to a situation where boiler oversizing is common and this gives poor efficiency and excessive fuel consumption. A difficulty has been getting an accurate heat loss determination for the premises, since this is not a calculation that is routinely carried out by architects.

Table 22: 'Rule of thumb' boiler sizing.

Building Type	Heating Requirement
Houses	$30–70W/m^2$
Hot water requirement	400–600W/house
Commercial buildings	$80–200W/m^2$

Using Wood Fuel to Heat a House

A typical three-bedroom semi uses around 17,000kWh of energy per annum to heat the house and provide hot water. One kWh is 3.6MJ, so around 60,000MJ is required to heat the house. Wood pellets have an energy content of about 18,000 to 20,000kJ/kg (c. 20MJ/kg). If we assume a 70 per cent efficient heating system, then around 86GJ (24,000kWh) of energy is required. This means around 4.5 tonnes of pellets are needed. Replacing coal with carbon neutral pellet heating will save around 6.24 tonnes of carbon dioxide per annum. A comparison between fuel price and carbon dioxide emissions is given in Table 23. Since fossil fuel prices are very volatile, they may vary significantly from those shown.

Table 23: A comparison between fuel price and carbon dioxide emissions.

Fuel	Price (p/kWh) (2008 prices)	CO_2 Emissions (kg CO_2/kWh)
Wood logs	0.0–5.0	0.03
Wood pellets (bulk)	3.0–3.5	0.03
Wood pellets (bagged)	5.0	0.03
Wood chips	1.5–3.0	0.03
Electricity	3.9–10.9	0.42*
Heating oil (condensing boiler)	4.2	0.27
LPG (condensing boiler)	4.7	0.23
Coal (anthracite grains)	3.0	0.29
Natural gas	2.8	0.19

Source: Based on modified Logpile data, available at www.nef.org.uk/logpile/pellets/cost.htm.

* Carbon dioxide from electricity will depend on the fossil fuel/nuclear/renewable generation mix on the electricity grid system at any given time.

Table 24: Wood fuel golden rules.

Top Tips for Biomass Stoves and Boilers	Things to Avoid with Biomass Stoves and Boilers
When fitting a wood fuel boiler, ensure the boiler is properly sized for the heat loss of the building. A proper building heat loss calculation should be performed before the boiler is selected.	Avoid an incorrect flue design. The installation of a properly sized flue of the correct height is one of the most important aspects of any boiler installation. Building Control Regulations require that an appropriate flue be installed.
Ensure that the building insulation matches or exceeds that required to satisfy the Building Control Regulations. Anticipate the requirements of the next round of Regulations.	Fuel quality is very important, as some boilers may be very sensitive to variations in fuel quality. In pellet boilers even small variations in the pellet quality could result in boiler problems.
Do consider fitting a pre-heat buffer tank to enhance the hot water storage. This will increase the system efficiency and provide a reservoir of hot water.	Pellet supply and quality are critically important. Sufficient dry storage is required to ensure that boilers do not run out of fuel. Dampness will cause pellets to degrade rapidly, making them useless.
If you are considering growing an energy crop, make sure there is a market for the product and remember SRC works on a two- or three-year start-up, so fuel needs to be sourced for the initial period until the supply is available.	When considering burning or using waste materials as an energy source (regardless of their origin), make early contact with the environmental authorities.

A slightly more accurate heat loss estimate for an existing property can be determined by reviewing the historic energy bills and heating pattern over several years, making suitable allowances for occupancy and structural alterations. From this, an idea of boiler fuel consumption can be made and adjusted for existing boiler efficiency. Domestic fossil fuel boilers around fifteen years old are probably operating at around 75 per cent efficiency. In a new building in the UK, the heating requirements are set down in the Building Regulations, which make reference to the calculation of the specific heat loss (in W/K or Watts per degree Kelvin) for the building. Fortunately, help is at hand in the form of the Seasonal Efficiency of Domestic Boilers in the UK (SEDBUK). Boiler sizing guides are available and the BRECSU Whole House Boiler Sizing Method (available through SEDBUK) and the HVCA/ CIBSE Domestic Heating Design Guide provide excellent sources for this information. Building Control Regulation reviews over the coming decade will require higher levels of insulation in new buildings, and this will mean that smaller boilers can be installed for the same heat demand. However, existing properties will not be affected.

SAFETY PROTECTION

Modern wood chip and pellet burners are equipped with safety devices designed to prevent burn back from the burner into the fuel supply system. Most safety systems are installed between the feed screw (from bulk storage) and the stoker screw (feeding the burner), and they are designed to interrupt the flow of fuel. A temperature sensor controls the fuel feed safety devices, so that in conditions of high temperature (or a fire outside the boiler), or even if there is a power failure, the throttle valve or shut off valve is closed automatically. A rotary pocket feeder can provide additional protection because it operates in such a way as to provide a seal between the stoker screw auger (from the bulk store) and the feed screw auger (feeding the burner). Pellets are transported to the burner by a star wheel feeder, which rotates in a leak-proof, cast iron cylinder. Other safety systems might include water sprinklers or airtight fuel storage containers. It is also good practice to install a smoke alarm and carbon monoxide detector in any room that houses a boiler or stove. Safety valves, which vent to atmosphere when the pressure in the system gets too high, should be piped so that they release the hot water and vapour at ground level, preferably outside the boiler house.

WOOD FUEL APPLIANCES: PLANNING PERMISSION AND BUILDING CONTROL REQUIREMENTS

Planning permission is generally not required for a wood-fuelled boiler and associated equipment, but in some circumstances it may be necessary for the fuel storage building and the flue. Individual installations will vary greatly and it is difficult to give definitive guidance, so check the details of proposed installations with the local planning office. Proposals to modify the heating systems of buildings that have been designated as 'listed' by the planning authority, or if they are in conservation areas will certainly need discussion with the local planning office. A planning application will require a fee and these will vary, but if the application is made after the installation is in place (retrospectively), the fee will be higher.

In the UK a Building Control Certificate must be acquired for all new heating systems or for any modifications to existing systems. It would be advisable to contact the local authority or council (Building Control department) for further advice on the application process. The issuing of Building Control Certificates will attract fees and experience has shown that these may vary between authorities. These certificates will form part of the Home Information Pack (HIP), which is to be mandatory in the UK.

BIOFUELS GLOSSARY

Air-dried This means that the fuel has been left to dry in air, usually under cover.

Anaerobic digestion The breakdown of plant and animal material without oxygen by bacteria.

Appliances In this context, these are boilers and stoves.

Auger A screw-type feeder used to transport wood chips and pellets between the storage containers and the boiler.

Bulk delivery Delivery of wood fuels in significant amounts by specially adapted delivery vehicles.

Calorific value The energy value of different fuels.

Chicken litter A mixture of wood shavings, chicken droppings and feathers.

Combined heat and power (CHP) A system that produces both heat and electricity at a high system efficiency.

Coppicing A method of encouraging vigorous growth from plants and trees by cutting them back in the early stages of their development.

Cyclone (flue gas clean up) A device that imparts a spin to the flue gas to cause the particulates to fall out as ash.

Demand Side Management (DSM) The point where electricity demand is turned off when it is most expensive to serve, or an alternative source of generation may be provided instead.

District heating A single (large) boiler is used to heat a number of individual properties.

Ethanol A form of alcohol that can be used to run cars and vehicles.

Energy Services Companies (ESCOs) Companies that provide a range of products, possibly including fuel or electricity, as well as services such as energy efficiency, energy management control techniques and renewables.

Excess air Air in excess of that required for complete combustion of the fuel.

Forced ventilation The boiler draught is created by a fan.

Halogenated organic compounds Halogenated compounds have been used for a variety of purposes for hundreds of different industrial processes over the last fifty years. However, they present a danger to human health and include such known toxins and potential carcinogens as dioxins, pesticides and PCBs.

Heat loss: The ability of the fabric of the building material of a property to retain the heat supplied from a space heating system. The heat load is the demand for heat from a building.

Heavy metals In this context they are a range of metals present in biological systems. Some of the better-known elements (e.g. vanadium, cadmium and mercury) are toxic in low concentrations.

Homogenized Of uniform size and shape and consistent properties.

Hydrostatic pressure test A test where the boiler is filled with water to normal operating pressure and checked for leaks.

continued overleaf

BIOFUELS GLOSSARY *continued*

Methane A compound of carbon and hydrogen that is also a very virulent greenhouse gas. It is a major component of natural gas, landfill gas and biogas.

MWe Electrical power.

MW_{th} Heating or thermal power.

Naturally ventilated The draught to draw the combustion gases out of the boiler is created by the flue without using fans.

Organic materials Materials derived from living matter, such as plants and animals.

Pathogens Harmful or poisonous bacteria or substances.

Primary air The principal air for combustion introduced into the furnace combustion zone with the fuel.

Secondary air Air that is introduced into the boiler gas path and burns out with any unburned fuel in the later stages of combustion.

Sustainable In the context of energy, energy that will recur or will not deplete.

Thermo chemical processes A mixture of heat and chemical processes.

Volatiles Gases and substances with a high calorific value that are released from a fuel on heating.

USEFUL BIOENERGY CONTACTS AND WEBSITES

There are a number of regulations covering health and safety that apply to wood heating systems in relation to their construction, operation and maintenance. The main regulations are:

- Construction (Design and Management) Regulations 1994
- Pressure Systems Safety Regulations 2000
- Health and Safety at Work Act 1974

Bioenergy Feedstock Information Network
http://bioenergy.ornl.gov/
Gives biomass information from the US Oak Ridge National Laboratory (ORNL).

Biomass Pyrolysis Network (PyNe)
www.pyne.co.uk/
An IEA organization coordinated at Aston University in Birmingham.

British Biogen
ww.britishbiogen.co.uk
16 Belgrave Square, London SW1X 8PQ
Tel: 020 7235 8474 Email: info@british-biogen.co.uk
The British trade association for the biomass industry. Membership includes all the major suppliers of pellet boilers and associated equipment in the UK.

Centre for Alternative Technology (CAT)
www.cat.org.uk
Machynlleth, Powys SY20 9AZ, Wales
Tel: 01654 705989 Fax: 01654 702782
Email: info@cat.org.uk
CAT offers information, advice, training, consultancy and demonstration of various sustainable technologies, including all wood-heating technologies.

USEFUL BIOENERGY CONTACTS AND WEBSITES *continued*

Combined Heat and Power Association (CHPA)
www.chpa.co.uk/
The CHPA promotes combined heat and power and district heating in the UK, including schemes using waste incineration.

Department of Food and Rural Affairs (DEFRA) – Energy Crops
www.defra.gov.uk/farm/crops/industrial/energy/index.htm
DEFRA is the UK ministry responsible for farming and agricultural policy. This is the link to the page on growing energy crops.
DEFRA is also the UK ministry responsible for environmental protection (see www.defra.gov.uk/environment/index.htm).

HETAS Ltd
www.hetas.co.uk
Orchard Business Park, Bishops Cleeve, Cheltenham, Gloucestershire GL52 7RS
Tel: 01242 673257 Fax: 01242 673463
An official UK body recognized by government to approve solid fuel domestic heating appliances, fuels and services. They cover boilers, cookers, open fires, stoves and room heaters. HETAS operates a competent persons registration scheme for heating engineers with approved skills in the installation and maintenance of solid fuel heating systems.

Log Pile
www.logpile.co.uk
NEF Renewables, The National Energy Centre, Davy Avenue, Knowlhill, Milton Keynes, Bedfordshire MK4 1AP
Tel: 01908 665555 Fax: 01908 665577

Smoke Control
To find out whether or not you live in a smoke control area, visit www.smokecontrolareas.co.uk.

For details of local chimney sweeps, contact:

National Association of Chimney Sweeps (NACS)
www.chimneyworks.co.uk
Unit 15, Emerald Way, Stone Business Park, Stone, Staffordshire ST15 0SR
Tel: 01785 811732 Fax: 01785 811712
Email: nacs@chimneyworks.co.uk

Sustainable Energy Ireland (SEI)
www.sei.ie
Glasnevin, Dublin 9, Ireland
Tel: 00 353 1 8369080 Fax: 00 353 1 8372848
Principal contact for biomass activities in the Republic of Ireland.

Talbott's Ltd
www.talbotts.co.uk
Drummond Road, Astonfields Industrial Estate, Stafford ST16 3HJ
Tel: 01785 213366 Fax: 01785 256418
UK manufacturer of biomass combustion equipment.

'Waste Incineration – Research Paper 02/34'
A briefing paper written in 2002 for UK Members of Parliament, available at
http://www1.eere.energy.gov/biomass/

CHAPTER 5

Wind Energy

On a United Kingdom basis, the wind energy resource already identified on land and offshore can entirely satisfy the current Government targets in respect of renewable energy (15 per cent electricity by 2015), with no detrimental effects on the transmission and distribution grid. Ireland also enjoys an excellent wind energy resource, which again can provide all the necessary energy to achieve its 2020 renewable energy electricity targets.

THE WINDY ISLES

The British Isles are blessed with the best wind energy resource in Europe and this gives the potential to generate the majority of our electricity requirements from onshore and offshore wind farms. Over the last two decades commercial utility-scale wind energy has become a reality demonstrating increased availability, larger machines and greater confidence in output. Wind energy presents an irresistible opportunity to deploy a proven resource that could significantly contribute to our national requirements for sustainable energy over the coming decades.

WIND RESOURCE

In areas where the Sun heats the Earth's surface, the land gets warmed and so does the air above it. The heating of the air causes the molecules to begin to move faster and faster and spread out, reducing the air density. The heated air in one of these regions rises relative to the cooler air and it becomes an area of low pressure. In air over the sea and in cooler regions, however, the air molecules move more slowly and remain closer together leading to a higher air density. The cold, denser air in these regions is heavier and pushes down on the Earth's surface and oceans to create high pressure areas. The difference in air temperature between the two areas also results in a difference in the air pressure and, if this is big enough, the air moves from the area of high pressure (the cool air) to the area of low pressure (the warmer air). This movement of air is what makes the wind blow and the larger the pressure difference, the faster the wind blows

As noted, the British Isles have a fantastic wind resource, both onshore and offshore: estimates indicate that, in theory at least, it is possible to obtain more than 1,000TWh of electricity each year from the wind, assuming that it is technically and socially feasible. This figure means very little to the average reader, but it implies that wind-generated electricity could satisfy a demand for electricity several times that currently experienced in the UK and Ireland in a year.

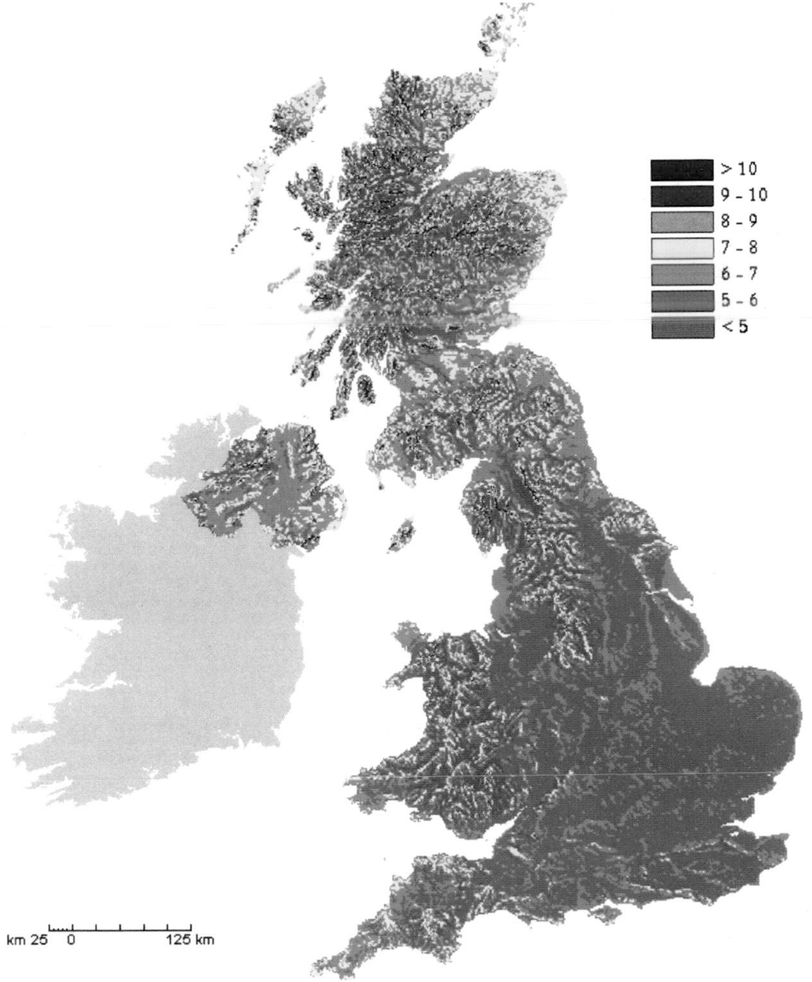

Figure 51: Average wind speed in m/s at 25m above ground level, across the UK. (Reproduced courtesy DBERR)

The reason these islands enjoy such a windy regime is that the Polar Maritime Air Masses (that is, the cooler high-pressure air masses in the region of the North Pole and its surrounding seas) travel over the Atlantic Ocean towards the warmer low-pressure regions further south. Most of the air mass originates in the Canadian Arctic or in the Greenland area. Eventually they reach the British Isles from the west or northwest after having swung around the western side of a depression (low pressure area). There are other air mass regimes that interact with the UK, but the Polar Maritime Air Masses are the most significant. The air mass atmospheric pressure, air temperature and humidity largely determines the energy content of the winds associated with these air movements.

In the UK and Ireland, locations near to the western coastline and those situated on hills or mountains are the windiest. Wind blowing onshore from the sea has less exposure to the drag and turbulence from the land. As the wind encounters mountains and hills, it will blow faster as the air mass is forced to rise over or around the obstruction. Wind speed also increases with altitude due to reduced friction between the layers of air. Wind is continuously variable on longer timescales, that is from season to season, but there will also be momentary gusts and lulls from minute to minute. For the British Isles, the summer months and daytime hours normally feature lower wind speeds compared to the winter and at night.

The strongest winds in the UK are associated with the passage of deep depressions (air masses of extremely low pressure) around or across the land. These are frequent during winter when the depressions are most active over the open ocean. The western coasts are exposed to the stronger winds at low altitudes, with Ireland affording protection to much of England. The most exposed areas of Great Britain and Wales are the coastal regions of Devon and Cornwall, where there are about fifteen days of gale a year. Moving inland, this reduces to fewer than five days a year. The highest gust recorded at a low-level site was 103 knots (118mph, 53.64m/s) at Gwennap Head, Cornwall, on 15 December 1979. The variation in wind speed, the frequency of gusts and, especially, the presence of low wind and the amount of time when there is a total absence of wind altogether are key parameters when considering the suitability of large- or small-scale wind power in an area.

THE HISTORY OF WIND ENERGY

Wind energy has been serving civilization for more than 4,000 years, propelling sailboats, grinding corn, dyes and spices, pumping water and sawing wood. The earliest designs were paddlewheels turning on a vertical axis and capable of taking the wind's energy from any direction. They required that the wind should be shielded on part of their rotation to provide the air pressure difference necessary to produce the rotational motion. These early designs were known as screened, clapper or differential rotation windmills, depending on the method employed to provide the pressure difference. Such machines have remained in use for several thousand years and are still around in some parts of the world today.

The more familiar horizontal axis windmill design emerged in Europe around the twelfth century. This presented the arms and sails of the windmill head-on, to face the direction from which the wind blows, and the oblique angle of the sail blades resulted in a rotation force perpendicular to the wind direction. In Mediterranean regions the horizontal axis windmills had triangular sails attached to the radial arms, whereas in northern Europe, notably the British Isles and the Netherlands, a four-bladed windmill design employing canvas sheets or sails was more common. The sails could be retracted or extended to provide an early form of power control. In some designs wooden slats or lattices were used; these could be opened or closed to provide the necessary speed regulation. The European windmills fell into two distinct classes: post mills and tower mills. These designs resulted from the need to move the mill so that it could face constantly into the wind as its direction veered. The post mill design used a wooden post (operated by the miller) to move the complete body of the windmill to face the wind. Tower mills featured a masonry or wooden tower that provided a tall support structure for the cap and blade assembly. Only this upper assembly rotated as wind direction changed. Long shafts and gearing were used to provide mechanical power. Tower mills frequently used small tail blades to drive the windmill around to face the wind automatically in response to changing wind direction. The introduction of steam power and the Industrial Revolution heralded the demise of the traditional windmill, although it is estimated that, in their heyday, there were in excess of 10,000 in Great Britain alone.

WIND TURBINE DESIGN

Modern windmills used specifically for electricity generation are referred to as wind turbines. This acknowledges the turbines, associated with steam and water, which were the traditional power generation technologies used to provide motive power. To produce significant amounts of electricity, a number of individual wind turbines of similar design are usually grouped in a wind farm. There is a huge variety of devices that can provide useful rotational motion from the energy of the wind. As we have already seen, the most popular can be classified as vertical axis or horizontal axis designs.

During the 1990s larger, utility-sized wind turbines (250kW and above) started to become commercially feasible. Danish and German designers already had horizontal axis wind turbines well advanced, so British designers sought a commercial edge by concentrating on a vertical axis concept. Vertical axis wind turbines

(VAWT) are not as efficient as the horizontal axis machines, but they do perform better in low wind speed situations. VAWTs are considered to be simpler to construct, they can be mounted close to the ground and have better turbulence-handling capabilities. They work well regardless of the wind direction, providing efficiencies of around 30 per cent. The most common designs are the Savonius and the Darrieus VAWTs. Power control for these machines is through reefing or arrowhead systems, which are fitted to the blade ends. Several large VAWTs were trialled at a wind test site near Caernarvon Bay in Wales. Unfortunately these machines had a troubled history, and the turbines and test site were eventually abandoned.

This marked the end of the British VAWT design effort and most commercial wind turbines since then have been of the horizontal axis wind turbine (HAWT) design. Most large-scale HAWTs and many smaller wind turbines have three blades, although the smaller machines can have many more. An early utility-scale (330kW) wind turbine, the MS-3, manufactured by a UK-based company, had two blades, but it never achieved commercial success. Small-scale wind turbines tend to be HAWT, but several micro VAWTs are now available. These might work more effectively than HAWTs in urban locations, where wind direction veers frequently and turbulence is a problem.

WIND TURBINE OUTPUT AND LOAD FACTOR

The load factor is used to compare the output from wind turbines at different locations. It is an expression of the percentage of time, in a year, for which the turbine will be generating at its rated output. The figures presented in Table 25 are typical and the actual load factor will depend on a variety of parameters, including availability, average wind speed at the site, the percentage time that the wind is at or above rated wind speed, the height of the wind turbine and the character of the surrounding land and buildings. These factors can be assessed at the time of the site survey to provide a reasonable estimate of the annual machine energy capture (or annual output in kWh). A load factor in the range of 30 to 40 per cent would be considered good performance.

Figure 52: At one time the UK landscape was littered with these corn-grinding windmills. This is a traditional cap and tower design at Ballycopeland, County Down, Northern Ireland.

Table 25: A comparison of wind turbine ratings, load factors and expected output.		
Wind Turbine Rating (kW)	**Load Factor (%)**	**Expected Output Electricity (kWh/annum)**
2.5	15–30	3,000–6,000
6	15–30	7,000–15,000
10	15–30	13,000–26,000
20	15–30	26,000–52,000
50	20–30	87,000–131,000
250	25–30	550,000–657,000
1,000 (1MW)	30–40	2,600,000–3,500,000
3,000 (3MW)	30–40	7,800,000–10,500,000

The estimated electricity output (P_e in kWh) is based on the turbine rated capacity (C in kW), the load factor (LF) and the number of hours in a year (8,760 hours). Output is then calculated from,

$$P_e = C \times LF \times 8{,}760$$

For a 2.5kW wind turbine this is 2.5 × 0.15 × 8,760 = 3,000kWh/annum.

For comparison, an average home will use around 4,000 to 5,000kWh (or 4,000 to 5,000 units, or 4 to 5 megawatt hours (MWh)) of electricity in a year. For the one megawatt (1MW) and three megawatt (3MW) wind turbines listed in the table above, the output energy is normally expressed as gigawatt hours or GWh, so for the 1MW turbine at 30 per cent load factor the output will be 2.6GWh, and for the 3MW wind turbine at 40 per cent load factor the output will be 10.5GWh. A typical wind farm built within the last five years will have around ten to twenty large wind turbines of 1MW to 3MW capacity. On a good site (with a mean annual wind speed of say, 8m/s), a wind farm could produce around 55 to 90GWh of electricity per annum, which is roughly equivalent to the annual demand from a small town of 10,000 to 20,000 homes.

As can also be seen from the above table, small-scale wind turbines may have load factors of between 15 and 20 per cent, since they tend to be mounted on lower towers and do not access the higher wind speed. The primary contributor to low load factor is average wind speed. Bad wind turbine availability (that is, turbine breakdowns) may be as a result of incorrect siting, poor design, or perhaps insufficiently robust fabrication or the turbine may require excessive maintenance.

A wind turbine has rotating components that, very much like a car, require greasing, oil changes and adjustment from time to time. This maintenance should be carried out in line with the manufacturer's instructions. For turbines under 50kW, the annual maintenance cost should be no more than that incurred with an average family car. Wind turbines are mechanical structures and may occasionally incur damage in extreme wind conditions.

WIND TURBINE COMPONENTS

Blades

Wind turbines collect the kinetic energy of the wind using aerodynamic blades connected to a rotor shaft. Generally small wind turbines (less than 100kW) have blades made from fibre glass reinforced polyester resins, while the larger wind farm machines use pre-impregnated glass reinforced epoxy resin moulded over wooden blade formers. Blades have an aerodynamic profile and this imparts a rotation to the hub as a result of the forces generated by the wind passing over their surfaces. Carbon fibre reinforced blades are becoming increasingly common due to their added strength and lower weight. In large wind turbines the blades are connected to the rotor hub through root studs.

Rotor Assembly

The rotor shaft is used to convey wind power to the electricity generator through a gearbox. Large machine rotors turn at speeds in the range between thirteen and thirty revolutions per minute (rpm), but this is too slow for electricity generation, so a gearbox is used to raise it to around 1,500 rpm. One problem is that the energy in the wind is

Figure 54: A 20kW wind turbine, such as this Jacobs model with a lattice tower, is suitable for medium-sized businesses or commercial applications, such as farms or industrial units. (Author/www.windturbine.net)

proportional to the cube of the wind speed. The result of this is that sudden rapid changes in wind speed, such as strong gusts, can send massive power surges through the rotor and gearbox: a doubling of the wind speed, for example, results in an eight-fold power surge. These surges need to be carefully controlled in order to protect the expensive rotor, gearbox and generator from the immense and potentially destructive forces created. The wind turbine structure itself can be damaged in very high wind speed conditions. In extreme winds, wind turbine rotors have shed blades; indeed in the early windmills described previously, sails were sacrificed to protect the windmill itself. Wind turbines have brakes that hold the rotor firmly when the turbine is parked during periods of light wind, or they can be applied when the turbine is shut down in high wind speed conditions.

Control of the rotor speed and power flux can be implemented by changing the pitch angle of the blades, or by designing the blades to produce aerodynamic stall at rated wind speed. Stall occurs when the wind speed increases to such an extent that the aerodynamic property of the blade disappears and the turning (or lift) force that produces the rotation no longer exists. This means that stall-regulated rotors run at constant speed (the rated wind speed), whereas variable blade pitch control allows rotors to turn at variable speeds to ensure lower loading on the drive train.

Tower and Nacelle

Various tower designs are used to support the nacelle, with tubular, galvanized steel towers (as used in street or road lighting) now very common. Lattice tower structures (similar to electricity pylons) are also used

Rated Wind Speed

Wind turbine output varies with wind speed and manufacturers provide power curves for their various machines. These show that, as the wind speed increases, there will be a point at which the wind turbine moves from rest to begin rotating: this is known as the start-up speed or cut-in speed. Although the rotor starts turning, the wind turbine may not begin to generate electricity immediately.

As the wind speed continues to rise, the wind turbine will start to produce an output (in kW) until the wind speed reaches the rated wind speed, when it will produce its full output. Some turbines spin freely in light winds but do not produce an output; this gives the rotor the momentum that will be useful if the wind becomes sufficiently strong to accelerate the rotor to commence generation.

Proven Patented Furling

In winds of above 12m/s or 25mph, the Proven's blades twist to limit power in response to high rpm

Low Speed Equals Durability

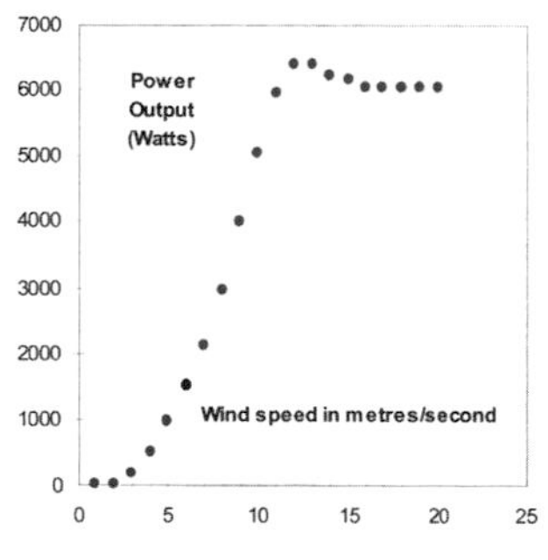

Marine Build Quality

All machines are manufactured with **galvanised** steel, **stainless** steel & **plastic** components

Technical Specification Sheet

MODEL	Proven 6 (6kW)
Cut In (m/s)[1]	2.5
Cut Out m/s)	None
Survival m/s)	70
Rated (m/s)	12
Rotor Type	Downwind, Self Regulating
No. of Blades	3
Blade Material	Glassthermoplastic Composite
Rotor Diameter(m)	5.5
Generator Type	Brushless, Direct Drive, Permanent Magnet
Battery charging	48V DC
Grid connect with *Windy Boy Inverter*	230Vac 50Hz or 240 Vac 60Hz
Direct Heating	ac
Rated RPM	200
Annual Output[2]	6,000-12,000 kWh
Head Weight (kg)	600
Mast Type	Tilt-up, tapered, self-supporting, no guy wires (Taller guyed towers also available on request)
Hub Height (m)	9 or 15
WT Found (m)	2.5x2.5x1 or 3x3x1.2
Winch Found (m)	1x1x1 or 1.5x1.5x1
Tower Weight (kg)	360 or 656
Mechanical Brake	Yes
Noise[3] @ 5m/s	45 dBA
Noise @ 20m/s	65 dBA
Rotor Thrust (kN)	10
Sample of commercial customers	British Telecom Scottish Youth Hostel Association British Rail Irish Lighthouse Authority UK Lighthouse Authority T-mobile Orange Shell Exploration Saudi Aramco

1 metre/second = 2.24 miles per hour=3.6kph
2 Output range is quoted to cover typical average wind speeds (annual). Lighter wind sites with typical 4.5m/s will produce lower end of range. Higher wind speed sites e.g. 6.5m/s average will produce upper end of range.
3 All readings taken with an ATP SL-25 dBA meter at the base of the tower at a height of 1.5m.
* A car passing 20m away @ approx 40 mph is 70-80dBA

Figure 53: The datasheet for the Proven 6kW wind turbine shows how the power output changes as the wind speed changes. This is known as the power curve for the wind turbine. The energy in the wind (P) is proportional to the wind speed (S) cubed, where P is in Watts (W) or kilowatts (kW), k is a constant and S is in metres per second (m/s). Written as a formula, this is: $P = k \times S^3$. (Illustration reproduced courtesy of Proven UK Ltd)

Rated Wind Speed *continued*

The rated wind speed will be different for each machine design, although many manufacturers of smaller wind turbines quote the rated wind speed as 12m/s. This is above the average wind speed for most sites, even in these islands, where a mean wind speed of 9m/s would be considered excellent. If the wind speed continues to rise above the rated wind speed, then the control system will take over to pitch (turn) the blades (in pitch regulated wind turbines) to an angle that provides the rated output. In stall-regulated machines, aerodynamic stall will ensure that the rotor speed remains constant (with a constant power output). As the wind speed drops and rises the control system will attempt to maintain the power output at a steady value, assuming the wind does not drop below rated wind speed. If it does, then the output will fall off in-line with the power output curve. The control system will respond to sudden gusts in the wind and will also try to smooth out variations in wind speed to provide as steady an output as possible.

In very high wind speed conditions when the control system is unable to prevent the power rising above the turbine's maximum rated value or when power surges may become dangerous, the control system will rotate or yaw the nacelle out of the wind. It will also apply the brake to ensure that the wind turbine does not experience destructive loading on the rotor components, gearbox or generator. This is the wind turbine cut-out or shutdown wind speed. Large wind turbines (over 50kW) are usually constructed to withstand winds up to 120mph.

Small-scale wind turbines should at least be rated to withstand wind speeds that average 35m/s (78mph) over a ten-minute period, without affecting their operation. In gusting conditions, the turbine and support structure should be designed to survive a severe gust of 50m/s (112mph) without suffering any damage that could lead to blades or other components falling from the turbine. Most manufacturers' instructions call for structural inspection of the turbines following the rare occurrences of gusts above 50m/s. Winds around 35m/s are more common and the turbines should be built to withstand them without any impairment in performance.

Table 26: Wind speed conversions.

mph	m/s	knots			Beaufort Scale
11.2	5	9.71	Typical cut-in wind speed	Gentle breeze	Force 3
26.9	12	23.33	Typical rated wind speed	Strong breeze	Force 6
44.8	20	38.86	Wind turbine generating under control	Gale	Force 8
56.0	25	48.57	Typical cut-out/shutdown wind speed	Storm	Force 10

One metre per second (m/s) = approximately 2.24 miles per hour (mph).
One knot (Kt) = 1.153 miles per hour (mph).

extensively (especially with wind turbines in the size range 10 to 50kW), although aesthetics and planning requirements have tended to favour the more expensive tapered tubular steel tower. Concrete pre-fabricated towers are available with heights of 90 to 100 metres for larger commercial wind turbines (larger than 100kW).

On large-scale wind farm machines, the tubular steel towers are around 4 metres across at ground level, tapering to 2.5 metres in diameter at the nacelle. In these machines there is enough space to house several electrical switchgear and control cabinets at the base of the turbine.

Nacelle is a term originating in the aero-engineering industry, referring to the structure, mounted on top of the tower, that houses the gearbox, generator, yaw control equipment (usually hydraulic) and some electrical switchgear. In wind farm turbines,

Figure 55: Lough Hill wind farm near Drumquin, Co. Tyrone, Northern Ireland, shown under construction during 2007. The wind farm has six Siemens 1.3MW wind turbines. (Photograph by Esler Crawford, courtesy of RES)

nacelles are typically around 5.5 metres long, 3.5 metres wide and about 2.5 metres high. They generally have an overall aerodynamic shape with a steel sub-frame and an outer skin made from reinforced plastic or glass fibre. The contents of the nacelle may be accessed via an internal ladder from the base. In very large turbines there may be a lift to the nacelle from ground level.

Tower height is vitally important and a modern 2MW turbine with a rotor diameter of 80 metres could have a tower height of 60 to 90 metres, depending on the annual mean wind speed and the topography of the surrounding area. Smaller machines for domestic or farm use (in the range 2.5 to 20kW) tend to be at much lower, perhaps 10 to 35 metres.

Gearbox

The gearbox is the workhorse of the wind turbine drive train. It converts the high-torque, low-speed input from the rotor and low speed shaft to the high-speed, low-torque output required for the generator. Wind turbine gearboxes are integral with the other drive components and are located in the nacelle housing. Gearboxes can be multi-stage helical, planetary or a combination of the two (hybrid). In some machines the rotor connects directly to the gearbox, while in other designs it connects to a low speed shaft and bearing arrangement to support the rotor. More recent wind turbine designs, even small-scale turbines, do not have gearboxes at all and the rotor connects directly to the electricity generator, which turns at the same speed as the rotor (generally between 18 and 38rpm). These direct-drive machines tend to be quieter than gearbox wind turbines since they completely avoid the gearbox mechanical noise (and associated losses). Modern wind turbines operate with noise levels just above those of a quiet bedroom.

Generator

Generators for wind turbines with gearboxes have traditionally been off-the-shelf units. These are readily available to the electricity industry from a range of suppliers, therefore keeping costs down. Most wind turbines currently operating have gearboxes.

Wind turbines designed to operate without a gearbox, such as the Enercon E-66/E-40 machines, require specially wound generators. As the rotor speed varies with the wind speed, the control of the power output and quality (in terms of voltage, current and frequency) is achieved through the use of power electronics. Direct-drive wind turbines with power electronics bring the advantages of better power quality, less stress on the rotor drive train, lower noise, lower cost and higher energy output.

Control Systems

A wind turbine is a complex dynamic machine and it requires a control system that can manage a range of rapidly varying parameters. Blade pitch control is used extensively to maintain optimum energy capture over a wide range of wind speeds. Wind turbines also employ a yaw control system to ensure that the turbine is always facing the prevailing wind direction. Large machines have anemometers for wind speed measurement and wind direction sensors that react to changes in wind direction by engaging the yaw drive to rotate the nacelle into the wind. The control system will also respond to gusts and rapid torque changes to minimize the loading on the drive train. As noted above, continuous control of the output power is required in order that the wind turbine does not disturb the grid frequency, or reduce the power quality of the electrical system to which it is connected.

From a grid system control point of view, wind farms are required to operate as if they are a single-connected entity. The wind farm will be treated by the system controllers as a single unit and it will operate totally unmanned, under fully automatic control with remote supervision. Individual turbine status can be continuously monitored with performance data capture and condition monitoring to assess the turbine's operational state. Wind turbine availability is the amount of time the turbine is available to generate, after allowing for lost hours through mechanical breakdowns and downtime due to maintenance. Latest-generation large-scale wind turbines achieve very high levels of availability (estimated at 98 per cent), with around just forty hours of maintenance per annum scheduled. Most machines are designed to operate for around 130,000 hours over their twenty-year lifetime. These breakthroughs are finding their way to the smaller wind turbines, which can benefit from control systems that provide better efficiency and maximized output.

Electricity Transformers

Transformers are needed to allow the electricity generated by the wind turbines to be exported to the grid network. They step-up the voltage from the wind turbine generator terminal voltage to the appropriate grid voltage. Wind turbine terminal voltages are designed to be the same as the industry standard voltages, for example three-phase 460 or 690 volts. The grid voltage level that accepts the electricity from the turbines varies, but it is generally in the range of 33,000 volts (33kV), 132kV or 110kV (Northern Ireland), depending on the size (in electrical power) of the wind farm. These transformers may be located in the nacelle or at the turbine base; indeed they may serve only one or a group of wind turbines in a wind farm, or they may be located in a central substation for the whole wind farm.

ENVIRONMENTAL IMPACTS OF WIND DEVELOPMENTS

Renewable energy sources offer the potential to generate meaningful amounts of electricity without the environmental problems associated with conventional fossil fuel sources (*see* Chapter 1). While wind energy shares these advantages, some argue that they come with environmental drawbacks, most significantly visual intrusion. Most of the impacts associated with wind, however, may be managed by careful selection and siting of the turbines, although this may also constrain the extent of the proposed development and consequently the available energy output. Apart from when the turbines are being constructed and installed, wind turbines generate electricity without the release of carbon dioxide, acid rain gases, particulates or other pollutants. A local wind resource can provide indemnity against the rise in price of conventional fuels and can also protect against the threat of their discontinued supply. A community or an individual farmer may benefit substantially from the payments made by wind farm developers for the siting of turbines on their land.

Figure 56: This Vestas V47 660kW wind turbine is satisfying a large portion of the Antrim Area Hospital's electricity requirement. (Author/www.vestas.com)

It is increasingly common to find large individual wind turbines supplying schools, farms, trade parks and hospitals. A typical wind farm may have ten or more machines with an output of several tens of megawatts. An example is the Callagheen wind farm in County Fermanagh, Northern Ireland, which has thirteen 1.3MW Siemens wind turbines, making a total of 16.9MW.

Wind turbines (and wind farms) require to be situated where the best wind regime is available. For large onshore wind farm developments this will be on the hills or mountain slopes in the countryside. Consequently such developments will be considered to have an environmental impact. Many of these areas are classified as Areas of Outstanding Natural Beauty (AONB) or they may carry similar protective status. Possible additional environmental impacts from these developments include interference on radio, television, microwave, radar and navigation signals (referred to as electromagnetic interference), noise and flora and fauna impacts.

From a landscape perspective, what we have come to think of as 'the countryside' has in fact changed dramatically over the last 2,000 years and very little of what remains pre-dates this. We are in a period of rapid climate change, which will have a potentially devastating effect on existing landscape character, and wind energy developments must be viewed in this context. Wind farms are ephemeral and reaction to their development, regardless of any careful environmental considerations, remains subjective.

Figure 57: This recent wind farm consists of thirteen Siemens turbines, each of 1.3MW. Callagheen wind farm is owned by Scottish Power and was constructed by Renewable Energy Systems Ltd. (Photograph by Esler Crawford, courtesy of RES)

Despite this, the factors that influence the visual impact of a wind development will be the turbine size, its design and colour, the number of turbines to be erected, the landscape character, the wind farm layout and the degree of potential distraction posed. Much work has been carried out to gain a better understanding of the less well-defined sociological factors that influence the visual impact of such developments. A number of large-scale wind farms have been completed in recent years, but some have attracted an increasingly ferocious opposition. Public attitude surveys have indicated that a small number of vociferous opponents or groups can influence decisions much more than a large number of supporters. Dedicated opponents conduct campaigns that can have a dramatic influence on the outcome of a proposed development. It is essential that local communities and the occupants of neighbouring properties are consulted at a very early stage, since these are the people who will be most affected by the proposal.

Planning Requirements

Different approaches to planning and wind energy have developed and guidance has been issued to planning authorities that recognizes the benefits and encourages the wider deployment of renewables. Wind turbines of all sizes currently require planning permission. For the United Kingdom planning application details are listed in the *Planning Policy Statement: Renewable Energy* (PPS 22). Scotland, Wales and Northern Ireland each has its own guiding planning legislation designed to encourage the development of renewable energy in those areas: respectively these are Scottish Planning Policy 6 (SPP6) 2007, Technical Advice Note 8 (TAN 8) and PPS 18.

Planning Policy Statement 22: Renewable Energy (PPS 22, 2004)

This document, issued by the former Office of the Deputy Prime Minister in 2004, suggests that planning applications for wind turbines should be accompanied by the following information:

- The overall economic and social benefits attributed to the scheme.
- Details of the potential impact of the project on nature conservation, including flora and fauna.
- Potential impact of the project on the built heritage and archaeology.
- Ground conditions including peat stability.
- Site drainage, sedimentation and hydrological effects.
- Size, scale and layout and the degree to which the project is visible over certain areas.
- Landscape character and visual impact issues, including access roads and electricity connections and substation.
- Local environmental impacts including noise, shadow flicker, electromagnetic interference.
- Adequacy of local access roads network to facilitate delivery of equipment and site construction.
- Information on cumulative effects due to other projects.
- Information on the location of quarries and borrow pits to be used during construction.
- Temporary and permanent storage and disposal sites.
- Decommissioning considerations.
- Wind turbine applications will be screened for the need to provide an Environmental Impact Assessment (EIA) where the development involves the installation of more than two turbines, or if the turbine hub height (or any other structure in the development) exceeds 15m. There are a number of issues specific to wind turbine developments and these need to be covered in an Environmental Statement. The Environmental Statement describes the significant environmental effects of construction and operation, identifying beneficial and adverse effects, together with relevant mitigation measures. It also considers issues such as the effect that the development might have on the market value of adjacent dwellings.

Specific turbine technical details:

- Rotor location – upwind or downwind.
- Hub height and rotor diameter.
- Speed(s) of rotation.
- Cut-in wind speeds.
- Measured background noise levels at specified properties and wind speeds.
- Predicted noise levels at specific properties closest to the wind turbine over the most critical range of wind speeds.
- A scale map showing the proposed wind turbine(s); the prevailing wind conditions; nearby existing development.
- Results of independent noise emission from the proposed wind turbine, including the sound power level. Predictions if it is a prototype.

In the Republic of Ireland the Department of Heritage, the Environment and Local Government (DHELG) has published similar guidance. The DHELG document *Wind Farm Planning Guidelines* is designed to assist with the planning of wind power projects in appropriate locations around Ireland. These guidelines, first published in 1996, have been revised and amended to account for changes in renewable energy policy and modern wind turbine technology evolution, to encourage wind energy at a localized planning level.

Communications and Navigation Equipment Impacts

Radio and television signal disruption, interference on communications systems, notably on microwave links, and also on radar and aircraft navigation equipment around airports, associated with wind turbines, is referred to as electromagnetic interference. Interference on these electrical signals does not imply any health impact, and the impact of wind turbines in this regard is no greater than that of a domestic electrical appliance. The extent of the electromagnetic interference can be influenced by the wind turbine design and by the materials used in construction of the blades and tower. Flat faceted towers (those with hard, sharp edges), for example, reflect electromagnetic signals more than rounded towers. Fibreglass or glass reinforced plastic (GRP) blades absorb electromagnetic radiation, while blades made of metal or with metallic components will reflect it. The most frequent interference occurs with television reception and this can usually be offset by installing new relay stations, or by providing cable television services to affected areas. Microwave line-of-sight disruption can occur and this may disturb emergency services' communication networks (police, fire and ambulance services). Most of these impacts can be mitigated by sensitive siting and early consultation with the appropriate groups.

Noise

Modern large-scale wind turbines are much quieter than their earlier forerunners, largely through improved design. The two sources of noise in wind turbines are, most significantly, the mechanical noise from the gearbox, hydraulic actuators and generator, and, to a lesser extent, the aerodynamic noise from the blades as they cut through the air. Gearbox and generator noise can be controlled by attention to design and through adequate soundproofing, which effectively makes the nacelle an acoustic enclosure. Aerodynamic noise, sometimes characterized as a 'swishing' sound, is due to a complicated set of interactions between the blade and the air flowing around it. Factors such as blade trailing-edge design, the interaction between the blades and the tower, the wind speed and blade speed, particularly the blade-tip speed, whether or not the blade is operating in stall conditions and the degree of turbulence in the region of the turbine, will all affect noise.

Wind Turbine Noise in Context

Source/Activity	Noise Level in dB(A)*
Threshold of pain	140
Jet aircraft at 250m	105
Pneumatic drill at 7m	95
Truck at 48km/h (30mph) at 100m	65
Busy general office	60
Car at 64km/h (40mph)	55
Wind farm at 350m	35–45
Quiet bedroom	20
Rural night-time background	20–40
Threshold of hearing	0

*dB(A): decibels (acoustically weighted) (Source: UK Department of the Environment 1993)

The World Health Organization (WHO) has issued the following guidelines on noise levels relevant to wind turbines:

'Occupational and Community Noise' (available at http://www.who.int/mediacentre/factsheets/fs258/en/)

'Guidelines for Community Noise' (available at http://www.ruidos.org/Noise/Comnoise-1.pdf)

In modern wind farms, the noise levels from turbines are so low for a carefully considered site that they would be drowned out by a nearby stream or by a moderate breeze in nearby trees and hedgerows (Source BWEA).

Wind Farm Noise

Since its publication, *The Assessment and Rating of Noise from Wind Farms*, ETSU-R-97, a report issued by the Working Group on Noise from Wind Turbines (September 1996), has been used to evaluate the noise from wind farms in the UK. When considering wind turbine noise, developers give consideration to wind speeds both at noise-sensitive locations and at the wind turbine site. When the wind speed is low, the turbines will not be generating and will therefore produce negligible noise. In medium to high wind speeds the background noise level at nearby properties due to the wind itself will generally be louder than the noise from the turbines. The most noise-sensitive periods occur when turbines are generating in low or medium winds and a noise-sensitive location is experiencing low wind-induced background noise. It is in these conditions that developers make their noise predictions. The models used to calculate noise assume a flat, hard ground with no buildings or other structures. This assumption produces 'worst case' noise assessments, since structures such as buildings, soft ground, trees and intervening hills would further reduce the actual noise from that predicted. When developers select turbines for a consented scheme, manufacturers are usually required to guarantee that the noise produced by their machines will not exceed the predicted levels, and to bear the financial cost of remedying any noise issues, should these arise.

In the assessment report, noise levels are set to safeguard the amenity at all dwellings. For quiet rural areas, such as those around a proposed wind farm site, these levels are described as follows: 'In low noise environments, the daytime level of the LA90, 10 min of the wind farm noise should be limited to an absolute level within the range of 35–40dBA.'

A differentiation is made between dwellings associated with a project, for example those belonging to a landowner, and those with no association to the project. For dwellings associated with a project, the assessment report recommends 'that lower fixed limits can be increased to 45dBA and that consideration should be given to increasing the permissible margin above background where the occupier of the property has some financial involvement in the wind farm'. Therefore, most proposals are usually assessed using the lower limits of 35–40dBA at the nearest dwellings and up to 45dBA at properties where the owners have an interest in the project, although noise levels in the UK are kept to well below this recommended level. It is important to note that these limits apply to noise levels outside the dwellings, as the assessment report is aimed to protect the amenity of areas used for relaxation and where a quiet environment is highly desirable. Noise levels inside a property will be approximately 10dBA less than those outside, even when a window is open. Noise concerns voiced at the planning stage rarely continue following the commissioning of turbines, and are generally thoroughly regulated through the use of appropriate planning conditions.

Aerodynamic noise tends to increase with faster rotation and designers try to keep rotor speed low to minimize this effect, especially in low wind speed conditions. Larger machines tend to be quieter than smaller turbines and in light wind conditions noise from wind turbines tends to be more audible. A wind farm at 350 metres distance would have a noise level of around 35 to 45 decibels (dB(A)). In practice, ground noise from grass, trees and other vegetation increases with wind speed and wind turbine noise gets lost into the background. In windy conditions, wind turbine noise is less likely to be a nuisance. It is also important that turbine developments are sited far enough away from inhabited dwellings to keep levels to a minimum. Noise from small-scale wind turbines can be more intrusive, since they will be sited much closer to (or even attached to) habited dwellings.

Shadow Flicker

Shadow flicker arises when sunlight falls on turbines, producing shadows that rotate or flicker across windows and buildings in a potentially distressing way. Shadow flicker can be largely eliminated by careful siting of the turbine. In extreme cases blinds or shades may be necessary for the periods during which the flicker arises.

ECONOMICS OF WIND POWER

The economics of large-scale wind energy developments, such as a multi-turbine wind farm, is outside the scope of this book. Any investment on this scale will require a proper business plan to be prepared with a full understanding of the costs, revenues and risks. With the rise in the number of wind farms there has been considerable interest from landowners who are keen to reap the potential rewards from the wind. Good sites for wind development are important: a site with 10 per cent higher annual mean wind speed than another, for example, produces more than 20 per cent additional energy. Central to a successful development is the inclusion of local communities. In most cases once a wind farm has been built, over 99 per cent of the land can still be used for farming. The base of the towers take up only a small amount of ground space and the blades of the turbine need to be a long way from the ground to access the higher wind speeds. This leaves the space underneath free to use. Turbines need to be widely spaced so that the wind speed picks up between each one. The British Wind Energy Association (BWEA) currently operates a free service for those who are interested in developing their land, or leasing part out to a developer for wind energy development. The BWEA circulates contact details, including the Ordinance Survey grid reference of the site, to their membership, who will then get in touch if they think the land would be suitable for a wind farm. Many landowners currently gain substantial financial benefits from hosting wind energy projects. For developers, an understanding of the return on investments through government support is necessary. Currently the UK is supporting large-scale wind energy through the Renewables Obligation (RO) (see Appendix I). In the Republic of Ireland the Irish Wind Energy Association (IWEA) is a useful source of information for potential developers and landowners, while the market support mechanism for renewables is the Renewable Energy Feed In Tariff (REFIT).

Wind turbines have the potential to impact the local community (both positively and negatively), and a growing number of wind farms have been developed in close cooperation with local community groups. This can involve groups buying into a single turbine or a wind farm. Some of the revenue from the electricity generated is returned to the local community, or it may be that the community is supplied with the electricity produced. Over half of the wind turbines in Denmark are operated in this way and similar cooperatives exist right across Europe. This approach can encourage a positive attitude to wind energy since the community benefits directly, and it makes the community much less hostile to commercial developers who may not be associated with the local area.

LARGE-SCALE WIND TURBINE DEVELOPMENTS

The current generation of large-scale onshore and offshore wind turbines have electricity outputs in the range of 3MW to 5MW, with rotors of the order of 100m diameter. The world's biggest wind turbine, the Enercon 6MW E-112, is located at Dardesheim in Sachsen-Anhalt, which aspires to be Germany's renewable energy village.

Offshore Wind

It was realized during the early 1970s that offshore wind energy had enormous potential, capable of providing many times the electricity requirement of the British Isles, and moving offshore avoids some of the planning complications experienced by onshore developments. The initial offshore wind farms to be constructed were Danish; the UK's first offshore development, near Blythe, Northumberland, was commissioned in December 2000. The UK encouraged offshore development by releasing tranches of projects in Rounds. Round One of the programme was unveiled in September 1999 with eighteen projects, each of up to thirty turbines. These sites were restricted to within 12 nautical miles from the coastline. In December 2003 the final results of Round Two were announced, with the rights to develop fifteen sites, totalling between 5.4 and 7.2 gigawatts (GW). The projects were awarded to ten companies or consortia.

The technology employed in offshore wind farms is very similar to those onshore, but with potentially larger turbines. A typical 3MW offshore wind turbine,

Figure 59: Construction of a wind farm at sea requires a stable operating platform (shown on the left of the picture), which facilitates turbine erection using a crane. A 'transition piece' between the wind turbine and the monopile (which penetrates the sea bottom) can be seen on the platform (in white). To the right are the turbine nacelles, ready to be loaded onto the platform. (Photograph courtesy of Airtricity)

with a height to the blade tip of around 120 metres (using 40m blades) will be mounted on a tower up to 80 metres tall. The Airtricity-owned Arklow Bank wind farm in the Irish Sea, for example, consists of seven General Electric (GE) 3.6-megawatt machines. Offshore wind farms are not space constrained to the same extent as onshore, so transporting large components such as blades is easier and the cost of offshore installation is much the same, regardless of the size of turbines.

Using larger machines gives higher energy yields, making them more cost effective. Offshore turbines are more likely to experience severe weather and must be able to withstand extreme conditions. As with onshore machines, for safety reasons the rotors will automatically turn out of the wind and slow down when wind speeds reach around 25m/s. The structural components of offshore turbines are specially designed and coated to protect them from corrosion by the salt in seawater. The UK and Ireland are presently identified as the best markets for offshore wind energy in the world, due to the favourable combination of wind resource, strong offshore regime and the extension of the facilitating legislation. The relevant legislation is the Renewables Obligation, which has been extended to provide 15 per cent electricity supplied from renewables by 2015, although it is expected to run to at least 2025.

OPPOSITE: Figure 58: The Arklow Bank 25MW wind farm, 10km out into the Irish Sea, has seven GE 3.6MW wind turbines. (Photograph courtesy of Airtricity)

First-generation offshore wind farms are sited in shallow waters so that their bases can be fixed to the seabed. This requires complicated and costly construction and it does not completely solve the problem of visual impact, since the rotating blades can still be seen from the shore. The ideal situation is to move further out to sea, into deeper water, although this brings bigger engineering challenges and requires a longer transmission line to get the electricity ashore. Further out at sea the wind supply is much stronger and more reliable, while obstructions such as trees, hills and buildings would not be a problem, although turbine stability and ensuring the assembly remains intact is fundamental.

Norsk Hydro has announced plans for the next generation of deep-sea offshore wind turbines, beginning with a demonstration floating turbine that, once constructed, will be anchored near the island of Karmøy, southwest of Norway. If the concept proves successful, floating turbines will be easier to make as most of the fabrication can be done on land and the completed machine then floated out to sea. Norsk Hydro's design is to use a concrete base as ballast for a 200-metre steel

Table 27: Small-scale turbine fully installed costs.

Wind Turbine Rating (kW)	Typical Fully Installed Costs (2008) (£1,000)	Cost per kW Installed (£1,000)
2.5kW	10–12	4.4
6kW	18–22	3.3
10kW	25–30	2.7
20kW	40–50	2.2
50kW	70–80	1.5

Source: Based on information from suppliers and various grant programmes.

Depending on location, the cost of connection to the local electricity network can be significant, sometimes as much as, if not greater than, the cost of the wind turbine.

tube, of which about half will stand above the surface of the water. The structure is kept secure in the North Sea, where waves can reach 30m in height, by three cables that loosely anchor the base to the seabed. The technology is similar to that used for offshore oil platforms. Engineers expect that the floating turbine could work in waters up to 700m deep and the prototype machine is expected to generate about 5MW of electricity. Clearly the engineering challenges of operating machines in the highly corrosive saltwater (and air) environment of the turbulent North Sea are immense. If all goes to plan, a small offshore wind park with about 200 turbines could be built by 2014, supplying electricity to coastal cities, or perhaps feeding into a pan-European offshore super grid, which some hope will be constructed off the coast of Europe to access the immense offshore wind resource.

A 5MW German RePower wind turbine has been installed in the deep waters of the North Sea, 44m below the surface and 24km off the east coast of Scotland, near the Beatrice Oil Field, since August 2006. Similar machines have been operating as prototypes since 2004 in Brunsbüttel, Schleswig-Holstein. Machines of this size require a 1,300m^3 concrete foundation, which is constructed from forty 24-metre long concrete piles and 180 tonnes of steel.

Wind conditions similar to those found offshore exist at the onshore DEWI-OCC test field in Cuxhaven, Germany, close to a North Sea dyke. A further two 5MW RePower wind turbines were erected there in December 2006. These have a rotor blade diameter of 126m, which sweeps an area of over 12,000m^2. The maximum electricity generation is achieved at around 14m/s, although significant output is generated even in a fresh breeze. The machine's rotors begin turning (cut-in speed) at around 3.5m/s, and the brakes are applied automatically at 32m/s (cut-out speed). In December 2007 the UK government announced that all UK homes could be powered by offshore wind farms by 2020. In a strong commitment to the offshore wind industry, it was announced that up to 7,000 wind turbines, two per mile of coast, could be installed to boost wind energy production. Larger prototype wind turbines of 6MW and 7MW have been constructed and a 10MW machine is under development in Germany.

SMALL-SCALE WIND ENERGY COSTS AND REVENUES

Small-scale wind energy projects (less than 25kW) and micro wind installations (less than 2.5kW) are less complex and can normally be undertaken by householders using registered installers. A small-scale wind turbine can generally make a simple payback in less than ten years at current electricity prices and taking into account export payments and environmental credits (ROCs). Payback is a useful means of assessing how financially viable the investment will be, but most accept that there are other added-value reasons for installing renewable energy technologies. In many cases these would take a higher priority than simple economic payback.

WIND TURBINE GRANTS

In the UK the uptake of small-scale wind systems for domestic, community and other buildings has been encouraged through government capital grant programmes (*see* Chapter 1), although no capital grant support for wind systems has been available in the Republic of Ireland. Due to the difficulty in estimating uptake in these schemes, the duration and

Figure 60: A Proven Energy wind turbine can supply medium-sized loads, such as 4- to 6-bedroom homes or large farms, with the provision to export to the grid. (Author/www.provenenergy.co.uk)

level of available grant funding is subject to change at any time. Grant allocation requires the use of certified equipment, which must be installed by an approved installer. It is, of course, possible to install non-approved equipment, but grants will not be payable and it is not recommended. (For contact details for these schemes, their accredited products and approved installers, *see* pages 25–27.)

INSTALLING A SMALL-SCALE WIND TURBINE

The first step in the decision to install a wind turbine is to ensure that sufficient wind is available at the site. There are a number of ways to determine what the actual average annual wind speed is for the chosen site. Large developments will require monitoring using an anemometer mast over a period of at least six months. The monitoring information can then be correlated to local meteorological office data and an estimate of the annual mean wind speed, wind direction and energy capture can be made. The predominant wind direction is important because steady, unidirectional wind is best. Wind speed and direction information can be provided in a number of formats, for example as a wind rose, which indicates the percentage of the annual hours during which the wind blows from a given direction at a given speed. Smaller-scale installations, especially those that may depend on a time-limited grant, will not have time for monitoring and they can draw from other available sources of information, such as postcode or map-based website wind speed data.

In the UK, the Department for Business, Enterprise and Regulatory Reform (BERR – formerly the DTI) website has a wind speed indicator, based on data from an air flow model (known as NOABL) that estimates the effect of topography on wind speed. The indicator provides only an initial estimate, since no allowance is made for the potentially significant effect of local, thermally driven winds, such as sea breezes or winds through mountains and valleys. The model was applied with one kilometre square resolution and takes no account of topography on a small scale, or local surface roughness, such as tall crops, stone walls or trees. These can have a considerable effect on the wind speed and energy capture. It is good practice to site the turbine at least ten times the tower height away from obstructions, if possible, particularly if they lie in the path of the prevailing wind. Each value stored in the database is the estimated average for one kilometre square at 10m, 25m or 45m above ground level. The database uses the Ordnance Survey (OS) grid system for Great Britain and the grid system of the Ordnance Survey of Northern Ireland, although the website will convert a UK postcode to an OS grid reference.

In Northern Ireland the Action Renewables website gives wind speeds at 30m, 75m and 100m above mean sea level for any Northern Ireland postcode. A similar indicator exists in the Republic of Ireland, but, since the country does not have a postcode system, the resolution is to one kilometre squares, based on the OS Irish Grid.

Having a good estimate of the annual mean wind speed is a prerequisite and the next stage is to select the wind turbine itself. Small-scale wind turbines can be used in a range of applications. They may be connected directly to an electrical or heating load (using immersion heaters); in isolated, off-grid locations they may be connected to a bank of batteries, or most commonly, they can be grid-connected. (For fuller discussion of battery banks and their connection to renewable energy systems, *see* Chapter 6.) Small-scale wind turbines come in a range of sizes: for an average domestic property or a farmhouse, the electricity output from a grid-connected 2.5kW or a 5kW turbine will make a significant reduction in the annual electricity bill.

The objective is always to get the wind turbine as high as possible (with the planners' permission). This must be balanced against the additional cost of the higher tower and the practical difficulties associated with cost and ease of lowering the turbine for maintenance and repair. Small-scale turbines are mounted on their own tower and good practice (from a safety perspective) suggests placing the tower base at least 1.5 times the tower height plus turbine height (to the tip of the blades) away from buildings and the nearest electricity line. From a turbulence and wind shadow point of view, the recommended spacing between turbines is between three and ten times the rotor diameter: for a wind turbine of 20m rotor diameter, the inter-turbine spacing in a wind farm will be between 60 to 200m, depending very much on the available space for the turbine layout and the prevailing wind direction. As it is better to keep the tower as far as possible away from buildings, this will usually mean that the connecting cable will be longer and larger (more expensive) and the associated cable losses will be higher (giving a lower output).

The connecting cable should be sized to ensure that the voltage drop at rated output along the cable is no greater than 4 per cent. Farmers and businesses may opt for larger machines, perhaps 20kW or even 50kW. This should only be justified on the basis of a business case decision, or a substantial grant, since these turbines will cost in the region of £40,000 to £80,000. Increasingly, refurbished ex-wind farm turbines of 100kW and above are becoming available. These large turbines are well suited for rural developments, public buildings, farms, hospitals, and community or business park applications.

Installing a wind turbine of any size is a specialist activity and requires the support of the manufacturer and a fully trained, accredited installer. Large machines can be installed in approximately a week and smaller turbines can be fully erected in a couple of days. Every turbine, regardless of the supplier or design, will require certain essential features:

- It should operate with low noise and vibration levels.
- It must be reliable.
- It should provide good value for money.
- It should be sited to have minimal environmental impact.
- It must comply with all necessary safety requirements, both structural and electrical.
- It is important to make sure that the installer is prepared to offer a maintenance contract for the turbine, since although the work is not complex, it is necessary and may require access to manufacturers' spares. It is good practice to seek quotations from at least two approved installers so that the fullest possible view of the costs and potential issues can be obtained. Installers should be prepared to provide examples of previous work that can be inspected by prospective customers.

PLANNING PERMISSION

Wind turbines of all sizes require planning permission, although there have been proposals to allow micro wind turbines to be installed on homes and other buildings on a similar basis to satellite dishes, as permitted development. When planning an installation, early contact with the local planning authority is recommended. Listed buildings and buildings in conservation areas will certainly require planning permission. It is recommended that the installation is discussed with neighbours, since they, along with the owner, will be the most immediately affected by the turbine.

Figure 61: Windy Boy wind turbine inverter, electrical isolator, generation meter and protection equipment mounted on a meter board. (Author/www.sma-america.com)

ELECTRICAL CONNECTION OF SMALL WIND TURBINES

Grid-connected wind turbines generate electricity that is synchronized with the public electricity system. Before connection, a Connection Agreement must be signed with the local distribution network operator (DNO) and the guidelines of *Engineering Recommendation G83/1* must be followed. This covers the operation of wind turbines that produce currents of up to 16 amps per phase, which is roughly equivalent to 5kW of wind turbine capacity. For larger systems than this, the recommendations of *Engineering Recommendation G59/1* should be followed. Grid-connected wind turbines will require an inverter, which converts the electricity from the wind turbine generator into fully synchronized electricity (of the correct voltage, phase and frequency) that is suitable for direct connection to the grid system. The inverter must comply with the requirements of G83/1 and a key feature is that it should protect against 'islanding' – that is, the turbine will disconnect electrically from the grid in the event of a loss of mains supply. The Electricity Supply Board (ESB) in the Republic of Ireland has similar requirements in respect of embedded generation plant. These are covered in detail in section CC12 of the *ESB Transmission System Connection Standards*, available from ESB National Grid. Another applicable document, also available from ESB, is the *Process for Connection of a Power Station to ESB's Transmission System*.

Appropriate metering will be required and this, along with details on how to access the value of spilled electricity (electricity generated by the wind turbine but not used in the premises), as well as its environmental value (LECs and ROCs), is described in Appendix I.

MICRO WIND TURBINES

Micro wind turbines are available that are suitable for attachment to a building's gable wall, using specially designed mounting brackets. These wind turbines, which are generally directly connected to the grid, produce outputs in the range 0.5kW to 2.5kW and they are predicted to generate several hundred kWh over the period of a year. Some micro wind turbines come with the option of connecting their output directly to immersion heaters, so that hot water can be stored whenever the turbine is generating, day or night.

The machines currently available have been slow to gain success in urban domestic markets for a number of reasons, in particular lower than predicted outputs. During 2006/7 the DIY retailer B&Q offered Windsave wind turbines in Ireland and the UK at around £1,800 to £2,000 fully installed, although this offer was subsequently withdrawn.

All wind turbines require a reasonable wind speed (around 6 to 8m/s) to produce a meaningful output and in most urban locations this is not available at rotor heights of 10 to 12m. Wind gusting and eddies are very frequent in built-up locations and these can significantly reduce the output of horizontal axis turbines. Planning authorities have been reluctant to allow micro wind turbines to rise much above the roofline and this leads to poor turbine performance. These wind turbines need to be mounted higher up (certainly above 10m) into the less disturbed airstream above buildings, and they require to be constructed with sufficient robustness to tolerate the stronger winds experienced at that height.

Horizontal axis wind turbines continually yaw into the wind to maximize their performance and in urban locations this can lead to a loss in energy capture, due to the constantly changing wind direction. Micro

LEFT: Figure 62: The Windsave 1.2kW small gable-mounted wind turbine was launched in 2006 to respond to a huge demand for domestic scale wind turbines. (Author/www.windsave.com)

BELOW: Figure 63: Two Swift 1.6kW Micro wind turbines mounted on the gables of the Berwick Housing Association residential home. (© Renewable Devices Swift Turbines Ltd., 2007)

Figure 64: Mounted on the Stadhuis, Den Haag, the 2.5kW Turby VAWT is designed for buildings in urban settings and can catch the wind from any direction. (Photograph courtesy of Dirk Sidler, Turby BV)

Table 28: Micro wind turbines.

Micro Wind Turbine	Rating (kW)	Type	Energy Capture (kWh/annum)	Cost (£) (approx)
Windsave	1.0	Horizontal axis	< 2,000	2,000
Swift	1.5	Horizontal axis	< 3,000	6,000
Turby	1.5	Vertical axis	< 3,000	12,000
Zephyr	1.0	Horizontal axis	< 2,000	–
Quiet Revolution	5.0	Vertical axis	< 10,000	33,000

Note: The cost data given in this table is for information only and is based on details given on the suppliers' websites and information sheets (2008). The Quiet Revolution wind turbine at 5kW is not a micro wind turbine as defined above, and it is included for comparison purposes only.

Table 29: Small-scale and micro wind turbine golden rules.

Top Tips when Installing a Wind Turbine	Things to Avoid in a Wind Turbine Installation
Before installation, get a good estimate of the energy output if you are installing the wind turbine for payback reasons. Pre-installation wind speed monitoring is recommended, but this can take too much time for those who wish to install a small-scale machine.	Do not overestimate the output energy available from a site. For larger machines (>6kW) a full business case should be prepared. Make sure the supplier/installer has not exaggerated the wind speed or energy output at the site.
Make sure the maintenance contract is negotiated as part of the initial purchase. It can be expensive to get a maintenance contract afterwards. Insist on a fully marked-up, as-installed electrical schematic for the wind turbine, as this will be essential for the DNO and for future maintenance/repair, should any problems arise. Make sure full operating instructions and guidance documents are provided.	Do not assume the process for claiming spill electricity (LECs or ROCs) is simple or straightforward ... it is not! This is possibly one of the most bureaucratic processes around. Proposals to award microgeneration double ROCs or the introduction of a feed-in tariff will significantly improve the economics.
Apply for planning permission and establish the electricity connection cost (from the DNO) as early as possible in the process.	Do not site a wind turbine too close to a building. It is good practice to site a turbine 1.5 times the height of the tower away from buildings and electricity lines.
Make sure a full assessment of the environmental impact is made and that the planners are involved from a very early stage. An Environmental Impact Assessment (EIA) and/or an Environmental Statement (ES) may be required. These are very important aspects of the project.	Do not assume that any wind turbine will suit any location.
Do try to get the wind turbine as high as possible above the ground. As a general rule, the higher the wind turbine tower, the higher the mean annual wind speed and the less the machine will be affected by building and ground turbulence – and the higher the output.	Do not install a wind turbine that is not fit for the wind regime at the site. This should be a prime consideration for the installer, and there are different classes of wind turbines. Machines designed for operation in continental Europe, China or the USA will probably not be suitable for the wind regime experienced during operation in the Scottish islands or northerly regions of the UK and Ireland.

wind turbines can also experience a significant reduction in operational lifetimes, due to the detrimental effect of the gusts and eddies on the electro-mechanical components, including the blades, gearbox, alternator and bearings. Vertical axis machines, which can take the wind from any direction, go some way to alleviate these problems.

Micro wind turbines that are fit for purpose need to be rapidly brought to the market with their performance validated by independent monitoring. Robustness of construction, external and internal noise and vibration, support and mounting integrity and nuisance need to be well understood. *Micro Wind Turbines in the Urban Environment* (2007), a study conducted by the BRE Trust (*see* page 132), illustrated that, in addition to the initial embodied carbon and efficiency of the turbine, the payback period is highly sensitive to local wind conditions, transport costs, maintenance requirements and lifetime of the turbine.

The technology exists to produce affordable, reliable micro wind turbines that can provide a meaningful contribution to offset the annual domestic electricity bill and help deliver zero carbon homes.

WIND TURBINE GLOSSARY

Accredited installer An installer who has undertaken specific training that is accredited or approved by the relevant grant awarding authority.

Aerodynamic A structure possessing a profile that gives rise to forces associated with an aircraft wing (lift and drag), for example a wind turbine blade.

Anemometer An instrument to measure wind speed.

Approved equipment Devices or renewable equipment that have been certified or accredited through approved standards or testing, and which are recognized by the relevant grant-awarding authority.

Availability Wind turbine availability is the amount of time the turbine is available to generate, after allowing for lost hours through mechanical breakdowns and downtime due to maintenance.

Darrieus rotor A number of aerofoils vertically mounted on a rotating shaft or framework. This design of wind turbine was patented in 1931 by Georges-Jean-Marie Darrieus, a French aeronautical engineer.

(DNO) Distribution Network Operator The local electricity network operator that handles enquiries about connections to the electricity grid.

Drive train The components of the wind turbine that convey the power from the rotor to the generator. This includes the high and low speed shafts, gearbox and generator.

Energy conservation programmes Loft and cavity wall insulation programmes and the introduction of 'A' rated energy efficient appliances.

Horizontal axis Having the axis of rotation in the horizontal plane (i.e. the axis is parallel to the ground). This means the blades will rotate in the vertical plane at right angles to the ground.

Kinetic energy Energy associated with movement.

Pitch angle Angle of the wind turbine blade relative to the plane of rotation. At a low pitch angle the blades will generate a high amount of lift or rotation forces in the rotor. At high pitch angles the blades will stall (stop).

Power quality The steadiness or constancy of a range of electrical parameters, such as voltage, frequency and reactive power.

Savonius rotor Used in some types of vertical axis wind turbines for converting the power of the wind into torque on a rotating shaft. They were invented by the Finnish engineer Sigurd J Savonius in 1922.

Supply regulations Define limits or quality standards for grid electricity supplies.

Utility scale Of sufficient size to be useful to a utility company, such as the energy supply or generation companies.

Vertical axis Having the axis of rotation in the vertical plane (upright). This means the blades will rotate in a plane parallel to the ground.

Yaw Movement in the horizontal plane (from side to side). Sometimes referred to as azimuth.

USEFUL WIND CONTACTS AND WEBSITES

British Wind Energy Association (BWEA)
www.BWEA.com
Renewable Energy House, 1 Aztec Row, Berners Road, London N1 0PW
Tel: 020 7689 1960 Fax: 020 7689 1969

Centre for Alternative Technology (CAT Centre)
www.cat.org.uk
Practical working examples of renewable power, energy conservation, building your own sustainable home, etc. in mid-Wales.

Ecotech Centre
www.ecotech.org.uk
Turbine Way, Swaffham, Norfolk PE37 7HT
Tours available up a wind turbine that has a viewing platform at the top.

Irish Wind Energy Association (IWEA)
www.iwea.com
IWEA is a national association for the wind industry in Ireland. The website is a resource for IWEA members and others interested in the promotion of wind energy in Ireland. It contains up-to-date information on all aspects of the work of the IWEA and its members.

Wind Farm News
www.windfarmnews.org.uk
Wind farms and renewable energy news in the British Isles, updated daily.

Wind speed maps
Online resource available to determine average wind speeds at a given location to the nearest square kilometre across the UK: www.berr.gov.uk/energy/sources/renewables/explained/wind/windspeed-database/

Postcode coverage for Northern Ireland:
www.actionrenewables.org/site/Windmap.asp

Wind speeds in Republic of Ireland
www.sei.ie/index.asp?locID=283&docID=-1
http://esb2.net.weblink.ie/sei/mappage.asp

Yes 2 Wind
www.yes2wind.com/links.html

Site produced by Friends of the Earth, Greenpeace and WWF, with the aim of providing information and resources for the public to support wind farm proposals locally.

BOOKS AND PUBLICATIONS

Gipe, P., *Wind Energy Basics: A Guide to Small and Micro Wind Turbines* (Chelsea Green Publishing, 1999)

Installing Small Wind-Powered Electricity Generating Systems: Guidance for Installers and Specifiers, Energy Efficiency Best Practice in Housing (Energy Saving Trust, 2004); available at www.energysavingtrust.org.uk/uploads/documents/housingbuildings/ce72.pdf

Noise from Wind Turbines: The Facts (British Wind Energy Association factsheet, 2000); available at www.bwea.com/pdf/noise.pdf

Phillips, R., and others, *Micro-Wind Turbines in Urban Environments: An Assessment* (BRE Trust, 2007)

Piggott, H., *Choosing Windpower*, 4th edn (Centre for Alternative Technology, 2006)

CHAPTER 6

Solar Photovoltaic (PV)

Ten photovoltaic (PV) panels can provide around 30 to 40 per cent of the electricity requirements for an average family home over the course of a year. If production costs can be reduced and the value of PV generated electricity increased, then PV could play a significant role in realizing the 'zero carbon home'.

ELECTRICITY FROM SUNSHINE

Anyone fortunate enough to have a south-facing (or close to south-facing) roof can benefit from installing a solar photovoltaic (or PV) system. They utilize the free and almost inexhaustible energy that flows from the Sun. With an average lifetime of thirty to thirty-five years, they will provide electricity, silently and without pollution, with minimal maintenance, and they could reduce your carbon footprint by 30 tonnes of carbon dioxide emission over their lifetime. More than 30 per cent of our energy usage takes place in buildings (although this includes heat as well as electricity) and installing a solar PV system is one of the ways householders and other building owners can move to a more sustainable future.

Solar PV systems use solar cells that produce electricity when a range of intensities of daylight or solar radiation fall on them. In the United Kingdom and Ireland there is a useful solar energy resource (*see* Chapter 2), although around 50 per cent of the solar radiation available is diffuse, that is, indirect sunshine. Electricity can be produced even on cloudy days or when there is no strong sunshine, although maximum output will occur with strongest, un-obscured sunlight. The electricity is generated within a solid-state semiconducting material, usually made entirely from pure silicon. It has no moving parts and can operate almost indefinitely without wearing out. Solar PV can also be used as a building material in its own right and can be fully integrated into the roof or facades of buildings using solar shingles, solar slates, solar glass laminates, as well as a variety of other design options.

Solar cells have a multitude of applications; they are used in situations where mains or grid electricity is unavailable or impractical and versatility off-grid is limitless. They power a wide range of devices including Earth-orbiting satellites and deep space probes, consumer electronic systems such as calculators and wristwatches, motorway signs and lighting, and inaccessible communications equipment, and they are frequently used to power water pumps in remote, Third World locations. More recently, primarily as a result of government grant support for the domestic and community sectors, increasing numbers of grid-connected solar PV systems have been installed. Solar PV panels are very much part of the UK government's microgeneration strategy (*see*

Chapter 1), potentially providing a useful contribution to the longer-term sustainable supply of electricity. PV panels can contribute significantly to the electricity demand of domestic homes, commercial premises and community buildings and they will be an integral part of any zero carbon home design.

PHOTOVOLTAIC CELLS, MODULES AND ARRAYS

Bulk silicon is produced in long, cylindrical ingots and these are sliced into fine wafers. Once the wafers have been polished, doped and had an anti-reflection coating and electrical contacts applied, they are referred to as cells. Individual cells are combined to create a PV laminate or module, each of which can generate 1.35 volts. Combining thirty-six laminates will give a PV panel with an output voltage of 48 volts. A typical PV panel is roughly two metres long by one metre wide. Using round or truncated 100cm^2 cells gives laminates that are about 1,100mm by 600mm.

For existing buildings, PV panels are secured on a support framework that is normally mounted on top of the roofing material (usually tiles or slates). The support framework is attached to the wooden roof frame or lathes. If there is no suitably unobstructed space available on the roof or the orientation is too far removed from south, then they can be ground- or wall-mounted, again using specially fabricated supporting frames. PV panels normally lie flush with or follow the roofline, but increasingly (especially in new buildings) they may form part of or the entire weatherproofing element of the roof, replacing conventional slates or tiles.

Individual panels are comprised of a sheet of tempered glass on the front and a polymer or anti-static Tedlar encapsulation on the back. They are best fixed at an angle, facing south, or they can be mounted on a tracking device that follows the Sun, allowing them to maximize the capture of sunlight. Tracking is extremely rare due to the additional cost of the associated drive equipment. A group of ten to twenty PV panels, known as a PV string, can provide enough power for an average household. The complete installation is referred to as a PV array. For large commercial or industrial applications, hundreds of strings can be interconnected to form a single, large PV system or array.

PV panels vary considerably in appearance; most are dark in colour and they cover an area of typically nine to eighteen square metres for a house, to several hundred square metres for a school or office building. Panels may be connected in series connection and

Figure 65: A 25kW Solarcentury PV array mounted on the ground in front of the Northern Ireland Electricity area office at Ballymena. (Photograph Courtesy Northern Ireland Electricity)

Solar PV Definitions

Maximum power point This is the maximum electrical power output (P_m) that a solar PV cell can deliver. Cells can operate over a range of voltages and currents and this can be represented by a power curve or I-V curve (see below). Each type of solar cell will have a different maximum power point and this will be specified on the I-V curve for the cell.

Peak watt (or watt peak) The output of a solar PV cell will depend on multiple factors and it is normally defined at a set of conditions. The peak watt (W_p) is the output power with solar irradiance at a defined set of conditions. The standard test conditions (STC) are: irradiance assumed to be 1,000W/m^2, solar reference spectrum Air Mass 1.5 and cell temperature 25°C. Sometimes this output may also be referred to as the kilowatt peak (kW_p), or in large systems the megawatt peak (MW_p) of the cell.

Solar cell conversion efficiency This frequently used term is a measure of how effectively the PV cell converts the incident solar radiation into electricity. The cell efficiency (η) is the maximum electrical power output (P_m), divided by the input light irradiance under standard test conditions and the surface area of the solar cell (A_c in m^2). It is expressed as a percentage and can be represented by the following formula:

$$\text{PV cell efficiency } (\eta) = P_m / (E \times A_c)$$

At solar noon on a clear March or September day, the solar radiation at the equator is around 1,000W/m^2. This means that the standard solar radiation (E), or the air mass 1.5 spectrum, as it is generally referred to, is 1,000W/m^2. So a solar cell of 100cm^2 at 14 per cent efficiency would produce around 1.4W. A 1m^2 module would produce 140W.

Cost-efficiency Whilst not strictly a defined term, it is useful to consider the concept of cost in terms of electrical output. For example, a high efficiency multi-junction solar cell (see below) manufactured from indium diselenide, produced in low volumes, may have an efficiency in excess of 30 per cent. The indium diselenide cell could cost as much as one hundred times that of a mass-produced amorphous silicon cell, which has a much lower efficiency of only 8 per cent, but it may produce only four times the electrical power.

parallel configuration to create an array with the desired peak voltage and current. Connecting panels in parallel will produce higher currents. Series connection gives a higher voltage. The electrical power output from these arrays is given in watts (W) or kilowatts (kW) and it is normally given on manufacturers' data sheets as the peak (kWp) or maximum value achievable in standard test conditions (STC). Since the output power changes constantly (and linearly) depending on the incident solar radiation (or insolation), then the total electricity delivered by the array over, for example, a year is a more useful indicator of performance. The electricity delivered (or output) is given as kilowatt hours (kWh).

The performance of a solar cell is measured in terms of its efficiency at turning sunlight into electricity. Only sunlight of certain energies will work efficiently to create electricity, and much of it is reflected or absorbed by the material that makes up the cell. Because of this, a typical commercial solar cell has an efficiency of 14 to 15 per cent – about one-sixth of the sunlight striking the cell will generate electricity. Low efficiencies mean that larger arrays are needed, and that means higher cost. Improving cell efficiencies while reducing the cost per cell is a key objective of the PV industry.

PV CELL DEVELOPMENT

The earliest discovery of the photovoltaic effect is credited to the French physicist Alexandre-Edmond Becquerel, who noticed in 1839 that light falling on the silver plates of a wet cell battery increased the output voltage at the terminals. The PV effect was subsequently noted in a solid in 1877, when two Cambridge scientists, W.G. Adams and R.E. Day, observed the variation in the electrical properties of selenium when they exposed it to sunlight.

An American, Charles Edgar Fritts, is credited with constructing the first selenium solar cell in 1883, which was in many ways similar to the silicon solar cells we employ today.

The efficiency of these early cells was very low (around 1 per cent) and it was not until the developments in semiconductors in the mid-1950s that higher efficiencies became a possibility. Daryl Chapin, Calvin Fuller and Gerald Pearson, scientists at the Bell Telephone Laboratories in Murray Hill, New Jersey, showed that germanium and silicon semiconductors doped with boron and phosphorus could produce conversion efficiencies of around 6 per cent. In the years that followed, PV cells were used in a range of applications, for example as a power source for radio amplifiers in the Bell Labs, and they were also employed in the US Vanguard 1 space satellite. Progress in reducing the cost and weight of the cells combined with increasing the efficiency resulted in the widespread use of PV in the emerging space programme. Perhaps one of the most evident applications for these devices has been on a wide variety of space probes and satellites, such as the International Space Station. Back on Earth the potential for PV cells was soon recognized and they began to appear in a wide range of power supply applications in inaccessible or remote locations where conventional power sources were inappropriate, impractical or very expensive.

The larger-scale deployment of utility or grid-sized arrays (megawatt-scale electricity generation) has been demonstrated at an increasing number of locations across Europe and America. The largest PV arrays in the world in 2008 are all in Germany: at the Erlasee Solar Park, near Arnstein (12MW), Pocking (10MW) and Mühlhausen (6.3MW).

There have been a number of proposals to construct and place solar power satellites (SPS) in high Earth orbits. Once the SPS has collected the solar energy, microwave power transmission will be used to beam it to a very large antenna on Earth. One such proposal is Powersat, which resembles a giant bicycle wheel 1 km across, orbiting at an altitude of 480km. By 2025 a network of Powersats, using the latest PV cells, could generate 20 gigawattts (GW) of electricity. The practicality of this idea has yet to be demonstrated.

HOW PV CELLS WORK

Light absorbent semiconductor materials, such as silicon, absorb photons of solar radiation that produce electrons and positive holes in the silicon through the photovoltaic effect. Different PV materials absorb light of various wavelengths and these are selected for the task they need to perform, that is whether they will be receiving light at the Earth's surface or, perhaps, out in space to power a spacecraft. Most currently available solar cells are manufactured as bulk materials processed to provide silicon wafers. Some photovoltaic materials are in the form of thin films, which are deposited on supporting substrates such as glass or plastic.

Despite the development of these new technologies and innovative materials, for the foreseeable future the vast majority of solar cells will be reliant on the use of silicon. A brief description of each of the current PV varieties is given below.

Monocrystalline and Polycrystalline Silicon Cells

The majority of solar PV cells currently produced are made from monocrystalline silicon. Monocrystalline silicon, which has a single continuous crystal lattice structure (or framework), is manufactured using the Czochralski process, where a single silicon seed is used to create a very uniform lattice from molten polycrystalline silicon.

The cost of pure monocrystalline PV modules has been a problem since they are made using the slow, intricate, expensive Czochralski process, which was designed originally to provide very pure, electronics grade, polycrystalline silicon. Cells produced using the Czochralski process have a distinctive circular or squared-off shape, which makes them physically larger per kW output with non-productive areas between cells.

Polycrystalline silicon consists of small, randomly aligned grains of monocrystalline silicon. Polycrystalline silicon cells have the appearance of chopped-up bits of silicon in suspension. Polycrystalline solar cells are cheaper and easier to make than monocrystalline cells, but they have a lower efficiency. Careful orientation of the grains during manufacture can reduce the efficiency

Photovoltaic Effect (PV Effect)

The following is a simplified explanation of how a PV cell works. Generally a silicon solar cell is a wafer of p-type silicon with a thin layer of n-type silicon applied to one side. N- and p-type silicon can be created by doping the silicon to produce the required variety and bringing them together creating p-n junctions. Pure silicon is doped with phosphorus to provide n-type silicon, which gives a surplus of free electrons, and doped with boron to produce p-type silicon, which has a deficit of free electrons (or a surplus of positive holes).

When light photons of sufficient energy enter the silicon cell near a p-n junction, they may excite and displace an electron, leaving a positive hole near the junction as a consequence. The energy required to do this is referred to as the band gap. The displaced electron is mobile and tends to migrate into the layer of n-type material and the hole tends to migrate to the p-type silicon. The electron travels to a current collector (a metal strip) on the front surface of the cell, generates a current in the external circuit and then reappears in the p-type silicon, where it recombines with one of the waiting holes. It is this flow of electrons around the circuit that produces the electricity from the solar cell.

If the energy of the incident photon is greater than the band gap, it will give rise to an electron-hole pair as described, but the excess energy is converted to heat. Photons with energy less than the band gap will pass through the cell without interacting. Some photons will also be reflected off the front surface of the cell. Other photons will be absorbed or blocked from reaching the junctions because of the current collectors on the cell front surface, which are visible as metallic strips.

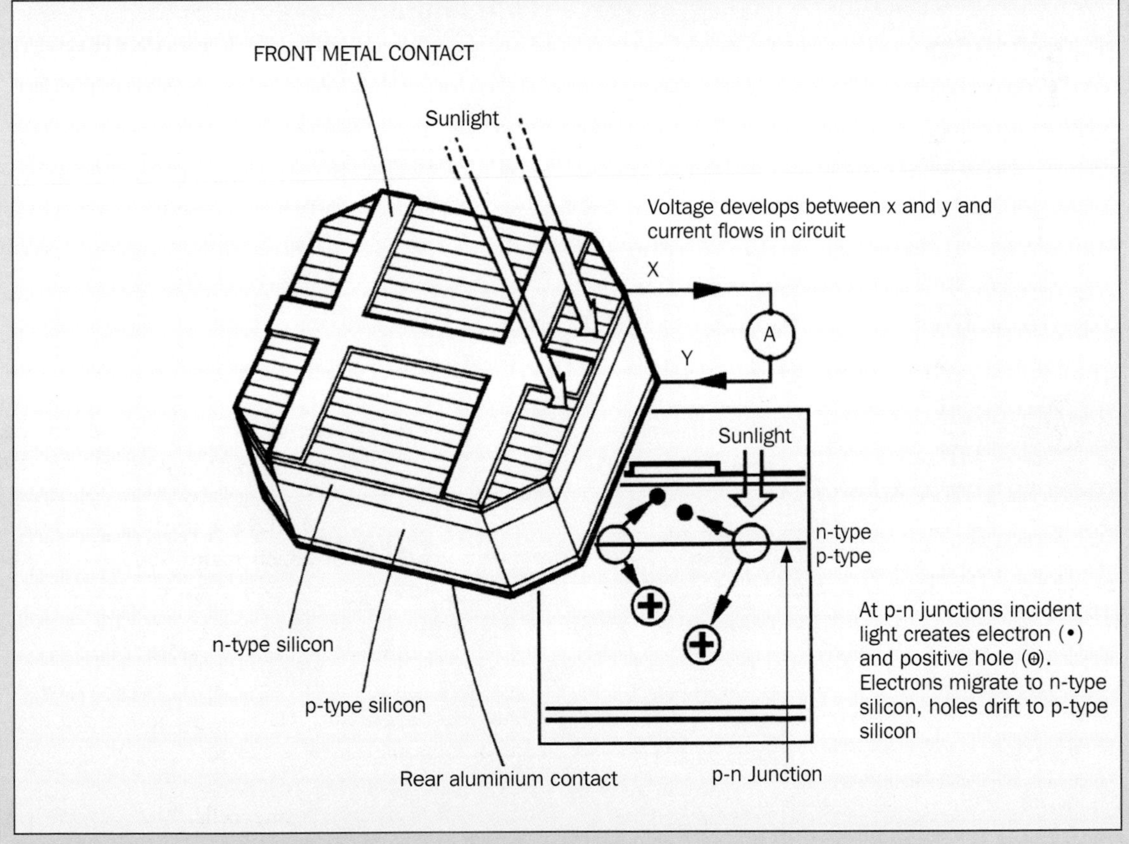

Figure 66: Principle of a silicon solar PV cell.

Figure 67: Close-up view of the crystalline structure of a Sanyo HIT hybrid module, showing silicon (blue) and surface electrical contacts (silver). (Author/www.sanyo.co.uk)

HIT PHOTOVOLTAIC MODULE
HIP-215NHE5

SANYO

The SANYO HIT (Heterojunction with Intrinsic Thin layer) solar cell is made of a thin mono crystalline silicon wafer surrounded by ultra-thin amorphous silicon layers. This product provides the industry's leading performance and value using state-of-the-art manufacturing techniques.

Benefit in Terms of Performance
High efficiency cell: 19.3%, **Module:** 17.2%
The HIT cell and module have the world's highest level of conversion efficiency in mass production.

High performance at high temperatures
Even at high temperatures, the HIT solar cell can maintain higher efficiency than a conventional crystalline silicon solar cell.

[Changes in generated power daytime]
Normalized Output power; 1.0; 0.5; Up 10%; Module temp.75°C; HIT; c-Si; Kobe 28.July.'02 Faced to South Tilt angle 30°; Time; 5 6 7 8 9 10 11 12 13 14 15 16 17 18 19

Environmental Friendly Solar Cell
More Clean Energy
HIT can generate more annual power output per unit area than other conventional crystalline silicon solar cells.

Special Features
SANYO HIT solar modules are 100% emission free, have no moving parts and produce no noise.
The dimensions of the HIT modules allow space-saving installation and achievement of maximum output power possible on given roof area.

Benefit in Terms of Quality
High quality in accordance with ISO 9001 and 14001 standards
HIT solar cell and modules are subject to strict inspections and measurements to ensure compliance with electrical, mechanical and visual criteria.

HIT Solar Cell Structure

p-type/i-type (Thin amorphous silicon layer); Front-side electrode; n; Rear-side electrode; i-type/n-type (Thin amorphous silicon layer); Thin mono crystalline silicon wafer

Development of HIT solar cell was supported in part by the New Energy and Industrial Technology Development Organization (NEDO).

Figure 68: The product specification for a Sanyo HIT HIP-215NHE5 hybrid module, which details the module efficiency as 17 per cent. The datasheet also describes the PV junctions in the cell. (Illustration reproduced courtesy Sanyo)

loss. Commercially manufactured polycrystalline modules (frequently referred to as semi-crystalline or multi-crystalline) can have efficiencies around 10 per cent. Polycrystalline cells have an advantage over their monocrystalline counterparts insofar as they can be shaped easily.

The use of solar grade silicon, lower in quality and cost, has recently become more widespread. The loss in efficiency has been minimal and simpler manufacturing processes have been employed. Also the production of silicon as long, thin strips or ribbons from polycrystalline material has helped to reduce costs. Innovative approaches have seen the introduction of new materials such as gallium arsenide and amorphous silicon. The use of light-concentrating devices and a variety of new techniques have improved the efficiency of PV cells with some reduction in cost.

Thin Film PV

Thin film silicon cells were developed in the 1990s, with cell efficiencies of around 15 per cent. Sometimes described as second-generation PV, thin film technology was designed to reduce the mass of material involved in the modules, with resulting lower costs, lighter cells and greater material flexibility. These thin film semiconductors are transparent, permitting the solar radiation to penetrate to the PV junctions.

Amorphous Silicon (a-Si)

In amorphous silicon (or a-Si), the silicon atoms have a more random orientation than in mono- or polycrystalline silicon. Amorphous silicon cells have the advantage that they are cheaper to produce than crystalline silicon and absorb light much better, permitting thinner, cheaper applications of the film. Cell efficiency is poor, however, sometimes as low as 6 per cent, and this can degrade through sunlight exposure over a period of months to around 4 per cent.

Amorphous silicon can be manufactured in a continuous, low temperature process and it can be applied, as a thin film, to a wide variety of substrate and support materials, such as plastic, glass and steel. Thin film PV lends itself to building integrated applications (BIPV) in the form of semi-transparent

Solar Radiation and PV Panel Orientation

We see solar radiation as white light, but in fact it is composed of a spectrum of wavelengths from infrared (longer wavelength than red) to ultraviolet (shorter wavelength than blue). PV cells admit full spectrum solar radiation, which passes through them to produce the PV effect. When radiation from the Sun strikes the Earth's atmosphere, it either passes straight through or it is scattered and becomes diffuse radiation. Direct radiation is the sunlight that appears to come directly from the Sun. It is surrounded by a region of less intense sunshine or diffuse radiation, providing diffuse lighting. On a cloudless, sunny day direct radiation can approach a power density of 1 kilowatt per square metre (1,000W/m^2), known as '1 Sun' for test purposes. In northern Europe this figure is closer to 900W/m^2 at ground level. At mid-northerly latitudes, it is normally assumed that around 50 per cent of the annual sunshine collected by a PV panel is direct sunshine and the balance is diffuse radiation. Direct radiation is more effective in solar concentrators but diffuse lighting provides most of our internal daylight in buildings and is sufficient to provide an electrical output from PV panels. Rooms with north-facing windows, or rooms that do not have direct solar exposure, are illuminated by diffuse radiation.

Solarimeters are instruments used to measure the solar energy incident on horizontal surfaces. More detailed measurements can determine the direct and diffuse components of the radiation and these can then be used to estimate the energy falling on tilted or vertical surfaces. There is considerable variation in the radiation levels between summer and winter in the British Isles. For example, in mid-July the radiation falling on a horizontal surface over the period of a day is around 4.8kWh/m^2. In January this figure falls to around one-tenth of its summer value, about 0.5kWh/m^2. The implications are that the solar resource is more useful in summer and, although there is a resource in winter, it is less intense and more difficult to collect.

To maximize the collection of this radiation, the PV panel surface should be tilted towards the Sun. This tilt angle is dependent on the latitude and the time of year. If the collector is tilted at the latitude angle, it will be perpendicular to the Sun's rays at midday in March and September. The tilt should be a little more horizontal to maximize summer collection and tilted towards the vertical to optimize winter collection. The tilt angle is not critical and for most applications the latitude angle will provide a good compromise. For northern latitudes, where solar heating systems will provide the best output in the summer months, it is recommended that the PV panel should be tilted at around 30 degrees to the horizontal for the UK and Ireland. If a tilted roof is not available or the orientation is outside the desired range, then wall or ground mounting the panels can be an alternative. Flat roofs, if they are sufficiently load bearing, can also support a frame-mounted PV panel or array.

Optimal orientation for PV panels is facing due south, but practical constraints may mean that panels can face any direction between south-east and south-west without serious degradation in performance. Orientations outside this, however, for example PV panels facing due west, east or north, will be extremely cost-inefficient and should not be installed. This means that, considering the random orientation of existing buildings, a large proportion of them would be suited for the installation of solar PV systems. (The information given in Chapter 2 concerning solar energy levels across the British Isles and the loss in effectiveness as a result of orientation and tilt applies similarly to PV.)

solar cells, which can be applied to windows. In this application, the film acts to reduce solar glare through the window and at the same time generate electricity. More recently, thermal processing techniques have permitted the application of crystal silicon directly to glass (CSG).

Compound Semiconductor PV Cells

Substances that are comprised of several different semiconducting materials (known as compound semiconductors) can produce the PV effect in a similar manner to silicon. These materials, such as copper indium diselenide (CIS), copper indium gallium selenide (CIGS), gallium arsenide (GaAs) and cadmium telluride (CdTe), can be manufactured as thin films. CIS semiconductor PV cells have reached efficiencies of 12 to 13 per cent.

Importantly, these materials have not demonstrated the performance degradation experienced in the a-Si cells, although thicker films are required and indium remains expensive. There are toxicity fears around the hydrogen selenide gas used in the manufacturing process, but overall development of these modules is encouraging. Modules with 8 per cent guaranteed efficiency have been available commercially since the mid-1990s. In CIGS technology, gallium is used to replace indium, since indium is relatively rare and therefore costly. Around 70 per cent of the world's indium is currently used to produce flat screen monitors and this could compromise solar cell applications.

Cadmium telluride (CdTe) is another compound semiconductor suitable for thin film applications. In the 1990s BP pioneered the manufacture of this compound of cadmium and tellurium using a relatively inexpensive and straightforward electroplating process. There is little evidence of the degradation with exposure to solar radiation seen in a-Si cells and efficiencies of around 10 per cent are possible.

Gallium arsenide (GaAs) is another compound semiconductor that can be used as a thin film PV material. These cells are very efficient and, unlike silicon PV cells, they are capable of high temperature operation without loss of performance. This makes them particularly suited for use in PV concentrator systems. Gallium arsenide cells are more expensive due to the manufacturing complexities and the scarcity of both gallium and arsenic. They are particularly suitable for spaceflight and military applications where performance, regardless of cost, is essential. They are both the most efficient (about 39 per cent) and the most expensive cells per unit area (c. £20/cm^2).

It is possible to improve the conversion efficiency of PV cells using stacked, tandem or multi-junction construction, where a number of junctions are layered on top of each other. Each layer can, in turn, extract energy from different wavelengths of the light as it passes through the cell.

PV is a Solar Battery

For most purposes, a PV cell can be considered electrically to behave as a solar-powered battery. Typically 100 square centimetres of silicon PV cell will generate around half a volt (0.5V) and deliver a current up to a maximum of 2.5 to 3 amperes in full sunlight. For the latitudes of the British Isles, full sunlight corresponds to around 1,000W/m^2. As a load (electrical devices, such as a lamp or cooker) is applied to a PV cell, the output characteristics (current and voltage) will change.

The power available is determined from the area under the I-V curve. As solar illumination reduces away from the maximum, the power curve stays the same shape but it moves to the left, so the output power available is reduced. The important parameters on this I-V curve are the open circuit voltage, the short circuit current and the maximum power point (see above). For PV cells the open circuit voltage is weakly dependent on the solar illumination, while the short circuit current reduces directly in proportion to the intensity of the incident solar radiation. The open circuit voltage decreases directly as the temperature increases. Broadly speaking, this means that the output power will increase as the strength of the incident solar radiation increases and the cell output power will fall off as its temperature rises.

The output from solar panels changes substantially with the variations in solar illumination, which can take place within short timescales. Most PV systems incorporate a maximum power point tracking system, which automatically varies the load seen by the PV cell in such a way that it is always operating at its maximum power point and so delivering maximum power to the load. To achieve higher voltages a number of PV panels can be connected in series. For example, if a PV panel is to be used to charge a 12 volt lead acid battery, then around thirty-six individual cells will be connected to ensure the charging voltage is usually above 13V, sufficient to charge a 12V battery even on cloudy days (i.e. 36 x 0.5V = 18V).

PV is a Solar Battery *continued*

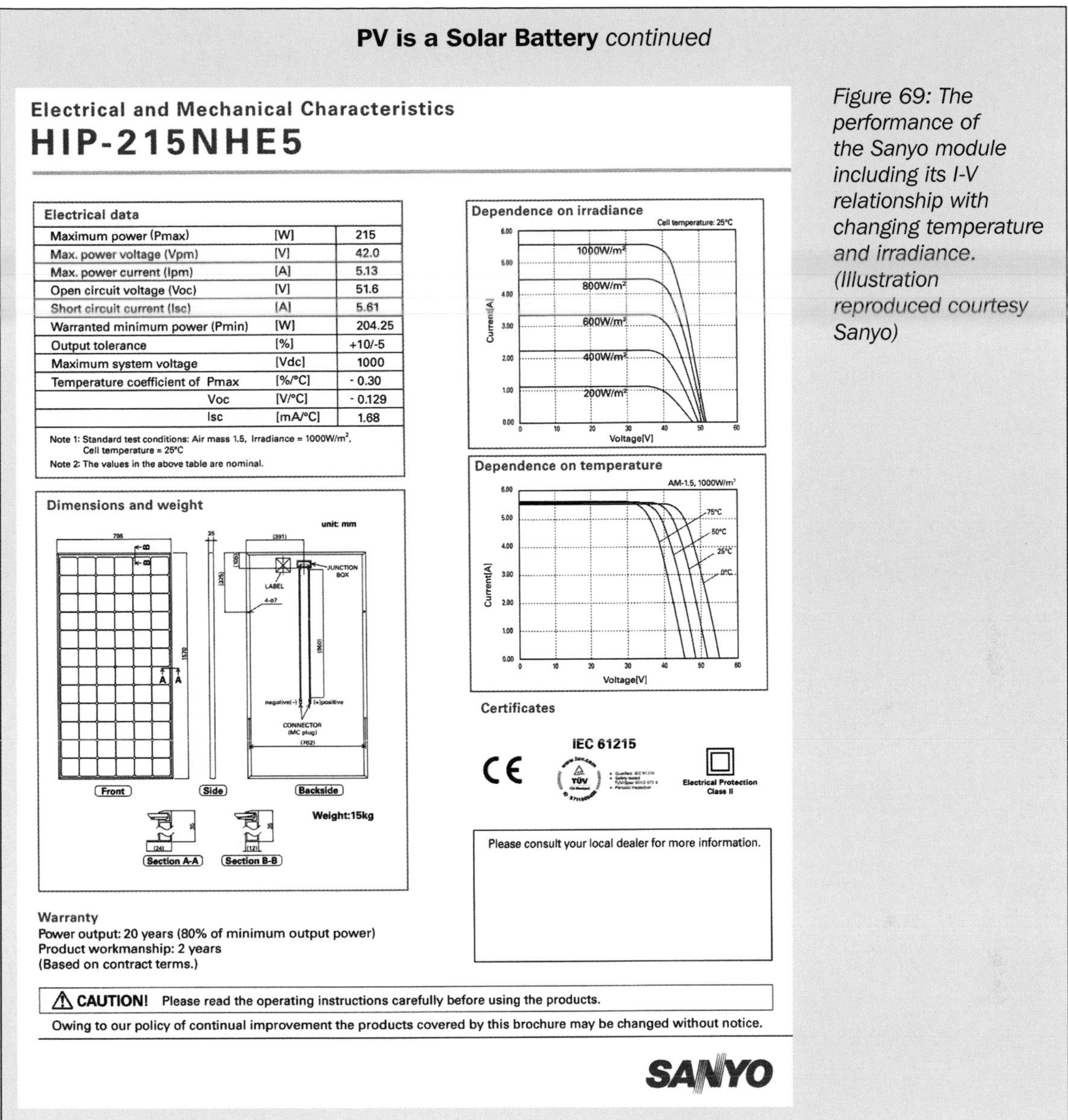

Electrical and Mechanical Characteristics

HIP-215NHE5

Electrical data		
Maximum power (Pmax)	[W]	215
Max. power voltage (Vpm)	[V]	42.0
Max. power current (Ipm)	[A]	5.13
Open circuit voltage (Voc)	[V]	51.6
Short circuit current (Isc)	[A]	5.61
Warranted minimum power (Pmin)	[W]	204.25
Output tolerance	[%]	+10/-5
Maximum system voltage	[Vdc]	1000
Temperature coefficient of Pmax	[%/°C]	- 0.30
Voc	[V/°C]	- 0.129
Isc	[mA/°C]	1.68

Note 1: Standard test conditions: Air mass 1.5, Irradiance = 1000W/m², Cell temperature = 25°C
Note 2: The values in the above table are nominal.

Certificates

IEC 61215

Electrical Protection Class II

Please consult your local dealer for more information.

Warranty
Power output: 20 years (80% of minimum output power)
Product workmanship: 2 years
(Based on contract terms.)

⚠ CAUTION! Please read the operating instructions carefully before using the products.

Owing to our policy of continual improvement the products covered by this brochure may be changed without notice.

SANYO

Figure 69: The performance of the Sanyo module including its I-V relationship with changing temperature and irradiance. (Illustration reproduced courtesy Sanyo)

PV CONCENTRATORS

Lenses and mirrors have been employed for some time to concentrate the amount of incoming solar radiation that falls onto solar PV cells. These can significantly increase the amount of electricity produced and can mean that fewer cells are required in an array. The concentration ratio is a measure of how effectively the concentrator is collecting the incident radiation, in relation to a reference flat plate array. Concentration ratios of two to several hundred are not unusual. High concentration ratios may be obtained with Sun tracking systems (or heliostats) to maximize the incident solar radiation. These mechanisms are often referred to as Heliostat Concentrator Photovoltaics (HCPV). They control the attitude of the solar collector through Sun position sensors that

Table 30: Solar cell material properties.			
PV Technology	**Advantages**	**Disadvantages**	**Efficiency**
Monocrystalline silicon	High efficiency. Widely used	Expensive and space-inefficient	14–16%
Polycrystalline silicon	Cheaper than monocrystalline Si. Can be shaped	Lower efficiency	10%
Amorphous silicon (a-Si)	Cheaper to produce than mono-Si and poly-Si. Can be used on various substrates	Low efficiency and degrades quickly.	5–6% falling to 4%
Copper indium diselenide ($CuInSe_2$ or 'CIS')	Relatively high efficiency and cheap manufacturing costs. Efficiency does not degrade	Indium is expensive and cell manufacture requires toxic hydrogen selenide gas	12.5%
Copper indium gallium selenide (CuInGaSe or 'CIGS')	As CIS but with gallium replacing indium. Higher cell voltage available and marginally higher efficiency	As for CIS	12–14.5%
Cadmium telluride (CdTe)	Relatively high efficiency and cheap manufacturing costs. Efficiency does not degrade	Cadmium is highly toxic	10%
Gallium arsenide (GaAs)	High efficiency. Tolerates high temperatures	Expensive to produce. Materials are scarce	Up to 39%

Table 31: Typical applications for PV stand-alone systems in the developing world.	
Agriculture	Water pumping and irrigation. Electric fencing for livestock and stock management.
Community	Water pumping, desalination and purification systems. Lighting for schools and other community buildings.
Education	Lighting, enabling studying, reading, income-producing activities and general increase in living standards. TV, radio and other small appliances. Water pumping.
Hospitals	Lighting for wards, electricity for operating theatres and staff quarters. Medical equipment. Refrigeration for vaccines. Communications (telephones, radio communications systems). Water pumping. Security lighting.
Miscellaneous	Lighting systems, to extend business hours and increase productivity. Power for small equipment, such as sewing machines, freezers, grain grinders, battery charging. Lighting, TV and radio in restaurants, stores and other facilities.

drive an elevation and azimuth motor control system. Such systems tend to be expensive and may use a lot of the collected energy when tracking. In some locations, even in the British Isles, despite the fact that concentrators are generally less efficient with diffuse radiation, it can be cost-effective to use fixed parabolic concentrating systems.

A different and more exotic concentrator mechanism is the use of fluorescent concentration. These devices consist of a block of fluorescent (or luminescent) dye between two sheets of plastic. The dye absorbs incident light and re-radiates it in a narrow band of wavelengths during fluorescence. The fluorescent light is mostly internally reflected and rebounds from three of the edges to impinge on the solar PV collector on the fourth side. These concentrators can be used with diffuse light, but work most effectively with direct solar radiation. They are unlikely to become commercially viable on a large scale.

Swiss researchers have been pioneering photoelectrochemical cells or PECs, which extract electrical energy from visible light. Although still very much in the domain of research, the objective is that PECs could eventually produce high efficiencies (about 10 per cent) and long-term stability with lower costs.

STAND-ALONE PV SYSTEMS

As noted previously, PV cells are frequently used to provide power in applications where it is inconvenient or expensive to use grid supplies. Increasingly these applications are non-domestic, ranging from PV powered microwave radio repeater stations on hilltops, to PV powered telephone kiosks. Smaller-scale applications charge batteries for boats and caravans, or power electric fences or even street lighting. It is now commonplace to see PV powered air quality monitors, weather stations and road signs, and in some solar water heating systems integrated PV panels are used to drive the circulation pumps (for illustration *see* Chapter 2).

In designing battery-based PV supported systems it is necessary to know:

- The size of the anticipated load (kWh, not kW);
- The daily, weekly, monthly and annual variations in the electrical load to be supplied;
- The available orientation and tilt angle for the PV array;
- The size of the battery, if one is needed;
- The predicted load cycle for the battery system (i.e. how often will it charge and discharge);
- The number of photovoltaic modules required;
- An assessment of the need for any back-up energy, allowing for load growth.

For developing countries, where very often the electricity grid is largely confined to the main urban areas, and where a substantial proportion of the rural population does not have access to most basic energy services, PV is widely regarded as the best and least expensive means of providing many of the services that are lacking. Table 31 lists a few typical applications for PV stand-alone systems in the developing world.

Figure 70: A pilot Green Column street lamp installation in Belfast. The 300 watt lamp is powered by lead acid batteries charged by PV and a small wind turbine. The unit is manufactured by Marlec Ltd. (Author/www.marlec.co.uk)

Designing an Off-Grid Stand Alone PV System

Computer programs have been developed by the various PV cell manufacturers to calculate the size and cost of the PV systems needed to support the proposed load. The following example shows the means of calculating the requirements for a PV-battery system to provide power for a 30W light for a phone booth in Belfast, which is at a latitude of 54.5° North.

To estimate the load cycle demand: For this example, assume that the 30 watt lighting (for a phone booth or advertising panel) is needed for sixteen hours a day in winter. Therefore, the total daily energy requirement is 30 x 16 = 480 watt hours per day. To allow for losses and cloudy weather apply a contingency factor of 1.5. This means 720 watt hours per day would be required.

To estimate the energy per surface area per day available from the Sun (in kWh/m^2/day): A PV panel tilted at the latitude angle gives the best energy capture averaged over the entire year, but circumstances may dictate that a different tilt must be used (roof pitch, for example). Changing the tilt of the PV panel can enhance the energy collection for a given season. In winter, energy capture can be enhanced by tilting the PV panel at a higher angle than the latitude angle. Although altering the tilt does affect the annual amount of energy collected, it is not as great a change as one might expect, usually only a small percentage. For the purposes of this example, assume that at the latitude of Belfast (54.5° North), the insolation for December, January and February averages 4.94 sun-hours a day, the summer months average 6.35, and the annual average is 5.79 sun-hours a day. The lowest figure should be used: 4.6 sun-hours a day in December. Note these insolation figures are available from local meteorological offices (see below).

To estimate the size of the PV array required: This is obtained by dividing the daily energy requirement by the sun-hours per day. For Belfast, the output of the panel is 720 divided by 4.6, giving an output of 156 watts.

Estimate the size of the battery: Batteries will last substantially longer if they are shallow-cycled, that is, discharged only by about 20 per cent of their capacity, rather than being deep-cycled daily. Deep discharge could mean that a battery is discharged by as much as 80 per cent of its capacity and a conservative design will save the deep cycling for occasional duty. This implies that the capacity of the battery should be about five times the daily load.

Considering the original design load (480 watt hours), apply a contingency factor of about 60 per cent to account for the efficiency of the battery discharge, the fact that only 80 per cent of the battery's capacity is available, and the loss in efficiency because photovoltaic systems rarely operate at the battery design temperature. The end result is that the battery design load is 480 x 1.6, or 768 watt hours. This is the daily energy drawn out of the battery, which is now multiplied by five to ensure 20 per cent daily discharge: 768 times five or 3,840 watt hours. This is the battery capacity, which is usually given in ampere-hours. The 3,840 must now be divided by the voltage of the system: 3,840 divided by 12 gives 320 ampere hours (320Ah). Notice that the battery's size is independent of the size of the panel or the solar resource, and is wholly dependent on the load and the assumptions around battery performance.

So our 30 watt PV-powered street lamp will require a 320 amp-hour battery and a solar panel of 156 watts output, which will generate at least 720 watt hours per day.

PV CONNECTED TO THE ELECTRICITY SYSTEM (DOMESTIC AND SMALL-SCALE)

Grid-connected small-scale and domestic PV systems have become increasingly common in recent years across the British Isles and Europe, particularly with the attraction of government capital support grants. PV panels are now available in a range of shapes and colours and they can be easily mounted onto existing roof structures or fully integrated within new roof designs. PV modules range in appearance from grey solar tiles (indistinguishable from normal roofing tiles at ground level), to transparent panels that can be used on conservatories or mounted on glass substrate (as windows) to provide shading as well as electricity. PV claddings attached to buildings can be employed as architectural features as well as to supply electricity. A prime consideration for many installations is the load-bearing capability of the roof structure. PV panels, as well as adding additional weight to the roof itself, are also subjected to uplift forces from the wind, which tries to lift them off the roof.

The electrical connections between panels are weatherproofed and brought inside the roof space to the inverter; the exterior wiring is largely invisible from the outside. On occasion it may be necessary to consider a structural roof redesign and replacement before an array can be attached.

Mains connection means that the grid provides a reserve when output from the PV panels is not available. As well as acting as a source of electricity in the absence of PV output, the grid can also accept spill or surplus electricity generation when the connected load cannot accept any further power from the PV modules. This is particularly useful when panels are producing electricity, with little or no available load. This can occur at times when the occupants are absent, perhaps on holiday or at work, or when there is little electrical demand from the premises. (For payment information in relation to electricity revenue from PV panels, see Appendix I.)

The grid can absorb PV power that is surplus to requirements very much in the manner of a giant battery that is being charged. This excess (or spilled) electricity flows to the grid and will automatically replace fossil fuel-generated electricity from the power stations. The PV electricity generated carries the additional benefit that it is supplied locally to customers and saves the electrical losses that occur in the grid transmission and distribution system (as the electricity flows through transformers, wires and cables from the power station). Grid-connected systems such as solar PV are described as network or grid embedded generation.

Figure 71: Inverter mounted in roof space with electrical isolators shown.

In grid-connected PV systems, the DC output voltage from an array requires to be converted to a voltage and frequency that can be accepted by the grid: 230 volts AC at 50 hertz in the UK and Ireland. This is done using a grid-commutated inverter, which makes sure there is synchronization between the PV electrical output and the electricity mains. The excess electricity not used within the building can be exported to the network and credited, with the agreement of the local electricity distribution network operator (DNO) or energy supply company. The rate paid for spilled electricity will very much depend on the spill payment offered by the energy supply company. Suitably certified and approved meters are installed to measure the amount of electricity generated by the system (generation meter) and spilled onto the grid (export meter).

Under some arrangements, payment is made for the spilled electricity as well as the environmental credit or greenness of the electricity generated by the PV system. The PV generation meter will register all the electricity generated by the panels. In some areas the energy

Figure 72: Northern Ireland Housing Executive flats in Belfast. This is a 51kW array of BP solar panels originally installed in 2000. (Photograph courtesy of Northern Ireland Housing Executive and Dr Steve Lo)

supply companies are considering the introduction of net metering systems. In this arrangement imported electricity is supplied and charged at the normal tariff rate for the installation, but exported electricity is deducted from the imported total and the installation is billed for the net balance between the import and exported electricity. It may be possible to have a net export of electricity with a payment for the exported balance from the energy supply company. This is analogous to the energy company's main import meter running backwards during export.

In the UK, connection of small-scale electricity generators such as PV arrays, which operate in parallel with the electricity distribution network, or grid, must satisfy Engineering Recommendation G83/1. In the Republic of Ireland similar connection requirements exist and details are available from the Electricity Supply Board (ESB). When installing a PV system, an application must be made to the local electricity company in order that it can assess the impact of the PV generation on the network. This application will be made through a special form, 'Application for Connection', which is an appendix to G83/1. Larger schemes may be required to meet the requirements of Engineering Recommendation G59. These recommendations specify the necessary electrical protection and determine the control required for variations in frequency and voltage. They also require the system to shut down safely in loss of grid situations. Copies of G83/1 and G59/1 may be purchased from the local electricity companies.

Once commissioned, the PV system is completely autonomous. It will come on when the appropriate level of daylight is available to power the system, and it will turn off at night when the light fades. If a shadow from a nearby structure, such as a

Figure 73: A retrofitted installation of ten 190Wp hybrid solar PV panels, manufactured by Sanyo, on a roof facing south-east. Peak output from the array is 1.9kW$_p$. Also visible is a thirty-tube Thermomax solar water heating panel on the roof facing south-west.

chimney, falls across the modules then a drop in output will result. The panels should be sited carefully, remembering that shadows will move with the position of the Sun, so account for its changing position through the day and from season to season. In winter the Sun will lie lower in the sky, which will give rise to long shadows. Most systems will produce typically four times as much electricity on a bright summer's day as on an average winter's day.

During operation all switches and circuit breakers in the system are switched on, and the inverter will be energized and functioning (a lamp indicates operation). During a power cut, grid-connected PV systems will stop operating and they will disconnect from the mains, as is required by G83 and G59. The system relies on the presence of the mains supply to function. It also ensures that any engineering work carried out to restore the electricity supply can be completed safely. Complete compliance with G83 or G59 is a condition of the connection agreement with the DNO. Once electricity supplies have been restored there will be no need for intervention from the user, since the system will automatically restart and resume generating, after a brief initialization period. It is worth reinforcing the point that during a power cut the grid-connected PV system will not provide a supply of electricity to replace the mains. Once the grid electricity supply is restored, the PV system will restart automatically.

MAINTENANCE OF PV SOLAR PANELS

From time to time, and depending on location, the panels may become dusty or dirty; in coastal locations they can sometimes be splattered with seagull droppings. Normally this will have a negligible effect on a panel's output, since most panels have a self-cleaning surface, which in effect means that the system is maintenance-free. It is recommended, however, that at least once per year, or better still every six months, the panels are cleaned gently with a damp cloth. If they are inaccessible it may be sufficient to hose them down with cold water. The aim is to maximize the

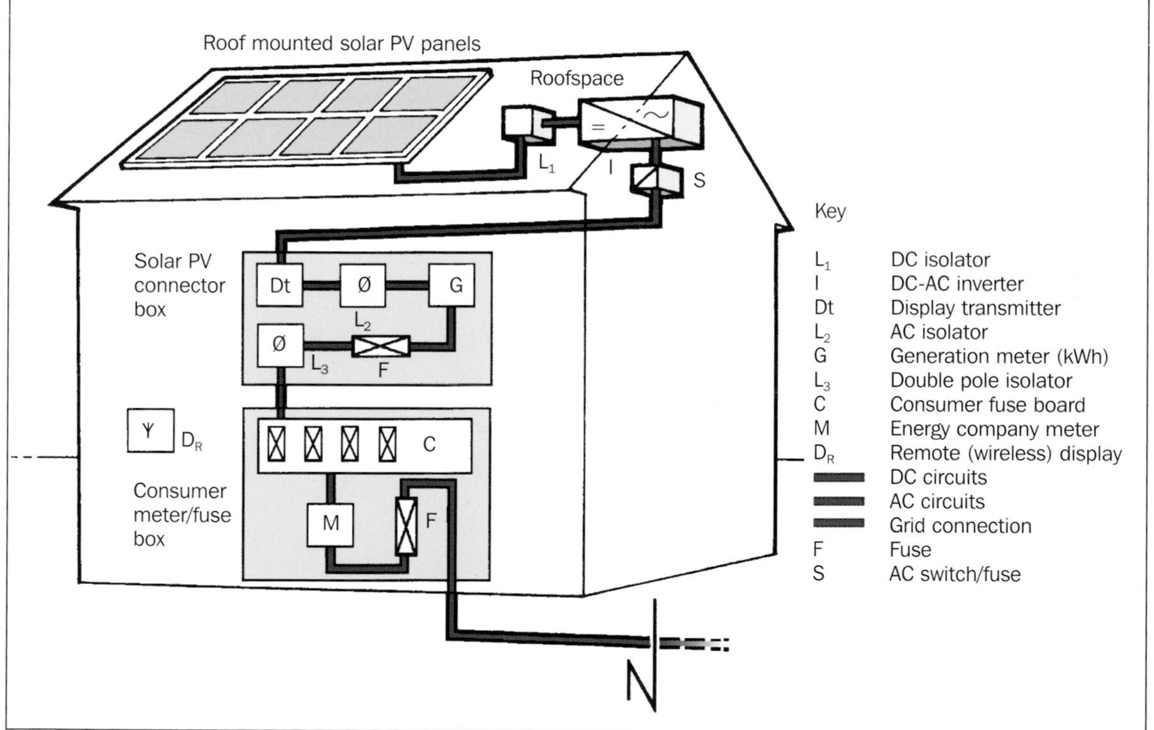

Figure 74: PV grid-connection schematic diagram.

amount of light incident on the panels. It is important to remove items such as leaves or litter that may have got stuck on any part or surface of the panels. This is because the cells underneath any such items will be permanently shaded and will be absorbing the energy from surrounding cells. This would result in irreparable damage to the panel and will impact on the performance of the array as a whole.

The inverter, which is normally mounted on a wall inside the roof space, requires airflow around it to provide cooling. It is important not to place items around or on top of it, to avoid overheating. In the past inverters have been one of the main reasons for system failures. Reliability has improved, although, as with all electrical devices, they can fail, and repairs must be carried out by a specialist. Most systems require an absolute minimum of attention and will function without incident for many years.

The PV modules are subject to the elements. In extreme situations, strong winds have been known to twist and bend the support framework. Water can corrode the framework and support structures, and on rare occasions it can penetrate the laminate coatings that are designed to protect the cells. In well-designed systems, the majority of these problems can be overcome by careful attention to material specifications, high quality control during manufacture and competent installation. Systems that involve battery storage will incur the additional maintenance requirements associated with keeping the batteries in good condition. Maintenance-free battery systems are available and they will run for longer periods without maintenance, but they are more expensive. In the final analysis, the maintenance costs of a PV system will be minimal in comparison with other energy systems. Installers, as well as providing suppliers' guarantees, will also be willing to provide support in the unlikely event of system problems.

PLANNING PERMISSION AND BUILDING CONTROL REQUIREMENTS FOR SMALL-SCALE INSTALLATIONS

Experience has indicated that it is difficult to generalize on the planning permission requirements for solar PV panels, but in most cases planning permission is not required for roof-mounted or wall-mounted solar panels, so long as they do not project beyond the existing roof surface plane by more than 150mm, and do not exceed the height of the highest part of the existing roof. Different conditions apply if the panels are to be erected elsewhere within the boundary of the property, if the building is listed, or if it is in a conservation area. However, this advice must be treated as guidance only and it is good practice to contact the local planning authority to clarify whether or not planning permission is required.

A requirement for solar PV panels that are to be installed on domestic premises is that a Building Control Certificate is granted. If you make any changes to a property, including installing a renewable technology, you may require Building Control approval. If approval is required, you will need to apply through your local authority or council Building Control Office. As with planning permission, early contact with Building Control is recommended. An application fee is required and this is determined by the nature of the work being undertaken.

THE ECONOMICS OF SMALL-SCALE PV SYSTEMS

The costs of any power system may be evaluated on the basis of the capital or up front costs, and thereafter the annual running costs incurred through maintenance and repairs. For a PV system these will include, as well as the cells, the cost of interconnection to form arrays, the support frame mounting system, possibly land and foundation costs (for large ground-mounted arrays), cabling, switching, metering, fuses and inverter(s), including the necessary protection devices as dictated by G83 or G59. Depending on the application, there may also be a cost for connecting to the grid, or upgrading the existing connection, or for batteries in stand-alone systems with their associated chargers, housings, fans and irrigation systems. The cost of actually installing the system (the wiring-up and fitting) is also part of these up front costs; because the modules are generally roof-mounted, scaffolding charges can make up a significant portion of the overall cost.

Most systems will have an associated display unit that can indicate, in real time, the PV system's output in kilowatts. This display is often remotely mounted or wireless, and can sometimes also display cumulated kWh, along with the associated carbon dioxide (CO_2) savings. These displays are a central, though not essential, part of the PV system, since they give a highly visible, real-time, indication of the output from the cells. They are also useful to inform whether or not the cells have stopped functioning. PV systems that have concentrators or tracking will also have the associated costs of these devices.

Figure 75: A wireless real-time display unit gives a continuous reading of the PV system output.

Grants for PV Systems

In England, Wales and Northern Ireland the uptake of PV systems for domestic, community and other buildings has been encouraged through a number of capital grant support programmes. Due to the difficulty in estimating uptake in these schemes, the duration and level of available grant funding is subject to change at any time. (For contact details for these schemes, their accredited products and approved installers, see Chapter 1.)

PV System Output

The output from a PV panel can vary rapidly and significantly from moment to moment, throughout the day. Assuming no cloud interference, panel output peaks around midday (see above). As a rule of thumb, at mid-northerly latitudes, a 1 kW_p PV panel will produce around 750kWh of electricity per year. The average electricity demand (not including electric heating, if fitted) in an average home is around 4,000kWh per year. This means that the panel could provide around 20 per cent of the electricity requirement. However, a problem is that, since the electricity is not stored in grid-connected systems, it must be used when it is generated to maximize its value, unless a favourable rate is paid for spilled electricity. In this case, it is better to have electricity available from the solar panel when it is being consumed in the building; this means that it offsets electricity that otherwise would have been taken from the grid at the domestic tariff rate (around 10 pence/kWh). All the renewable electricity generated by PV panels in the UK will be attributed Renewable Obligation Certificates (ROCs) and Climate Change Levy Exemption Certificates (LECS) in lieu of its environmental value. These are further described in Appendix I. PV systems will only generate electricity during the hours of daylight, so if it is not used then the electricity will spill to the grid. The value is the export electricity price paid for spilled electricity plus the value obtained from sale of the ROCs and LECs for all the electricity generated by the PV panels. Some energy supply companies will enter into an agreement with generators to purchase the spill electricity and the ROCs and LECs produced by the system. If net metering is available, then the value of the exported electricity will be the same as the imported electricity.

Most people ask about the payback against the PV investment (*see* Chapter 1). As an example, consider a 2kW_p PV system that costs around £12,600 installed (PV costs around £6,000 per kW_p installed plus VAT at 5 per cent). The grant will pay out around £3,000 per kW_p, so the installation costs £6,600 after grant. The 2kW_p system will generate around £150 savings per annum, assuming all the electricity is used when generated. The ROCs and LECs income contributes around £80 per annum (based roughly on the 2008 value), so a total income of £230 per annum is received. This gives a simple payback of roughly 28 years; this must be less than the expected lifetime of the system to be financially justified. This assumes the price of electricity rises in line with interest rates, but if the price of electricity increases ahead of this then clearly

Table 32: PV panel output and approximate cost.

PV Panel Rating (kW_p)	Approximate Installed Cost[1] (£)	Expected Output Range[2] (kWh)	Revenue Value of Output[3] (£/pa)
1.0	6,000	700–800	110–120
2.0	13,000	1,400–1,600	180–240
5.0	30,000	3,500–4,000	510–560
10.0	55,000	7,000–8,000	980–1,120

Notes:

1 The UK government has issued grants for PV systems under the DTI Major PV Demonstration Programme (2003), the Low Carbon Building Programme (LCBP, 2004) and the Reconnect Programme (NI only, 2006). Grants were payable up to the lesser of 50 per cent of installed cost or £3,000 per kW. Under the LCBP a grant of £2,000 per kW was available. In Northern Ireland the grant was further topped up by Northern Ireland Electricity, who offered an additional 15 per cent to bring the total available PV grant to 65 per cent under Reconnect. Costs based on grant programme quotations and actual costs quoted by installers.

2 Output kWh figures based on DTI/BRE PV in Buildings Field Trials programme and assume an output of 700–800kWh/kW_p per annum.

3 Revenue calculated on the basis, 10p/kWh displaced electricity plus £40/MWh for ROCs. One ROC awarded for 1,000kWh (= 1MWh).

the payback time will fall. Simple payback is a crude way to estimate the true value of a solar panel because the cost of carbon is ignored. A 1kW_p solar panel will save around 400kg/kW_p per annum of CO_2, as well as other pollutants associated with fossil fuel generation. Some energy companies have started to offer higher payments for exported electricity. In January 2008, for example, Scottish and Southern Energy (SSE) introduced a payment of 18p/kWh for units exported. This deal, known as Solar Energyplus, includes the ROC value and the company will fit an export meter.

Once installed, PV systems have no fuel costs and no moving parts. Consequently the running costs are very low and certainly less than other renewable energy systems, such as heat pumps or biomass boilers. The economic benefits from the cells arise from the displaced imported electricity otherwise drawn from the grid (now generated by the modules), as well as from the surplus electricity exported to the grid, supplemented by a payment for the environmental value of the electricity generated by the PV. In 2006 the DTI's Renewables Innovation Review estimated that solar PV could become cost-competitive with other forms of electricity generation by 2020 to 2030. To be most cost-effective, solar panels should be incorporated into buildings during construction, as this is cheaper than retrofitting. Building integration will increase the economic appeal of systems and the range of attractive applications.

ENVIRONMENTAL IMPACT OF PV SYSTEMS

Whereas, under normal operating conditions, PV systems emit no harmful gaseous or liquid pollutants, in a fire it is possible that CIS or CdTe modules could release small amounts of toxic materials into the environment. As with some renewable energy systems, most notably wind, PV panels can have a visual impact, although this is not necessarily negative. They are normally mounted on roofs or on the side of buildings, which gives them a high visibility that some people may object to on aesthetic grounds. Significant work has been undertaken in an attempt to produce cells that blend into a conventional roof's appearance. An example is the Solarcentury C21 system, which can be fully integrated into a roof and has the appearance of conventional roof tiles. The C21e version produces PV electricity, while the C21t variant is a solar thermal module with similar aesthetics and can produce hot water. The challenge for PV manufacturers is to design panels that fully integrate, cost-effectively, within conventional building designs and construction techniques.

PV cells have a long operating lifetime, but eventually it will be necessary to disposed of or recycle them. Silicon is not a significant problem, but methods to safely recover and recycle any toxic materials used during manufacture will be necessary. Installing $2kW_p$ of PV panels on an average domestic home will save approximately 800kg of carbon dioxide emissions per year, based on an annual output of around 2,000kWh and a saving of 0.4kg of CO_2 per kWh (against fossil fuel generated electricity) – totalling around 30 tonnes over a system's lifetime.

Energy payback (as opposed to cost payback) from PV systems has been controversial, with some suggesting that the systems do not pay back the energy consumed in their manufacture and installation. When a lifecycle analysis of the manufacture of PV cells is undertaken it is estimated that the energy payback for PV modules (excluding the other components of a PV system) ranges from about 2.1 years for crystalline silicon to 1.2 years for amorphous silicon. A typical PV cell has an expected operational lifetime of around thirty-five to forty years, giving an energy return ratio in excess of thirty. It is generally considered that energy and economic paybacks will reduce substantively in the future.

LARGER-SCALE AND UTILITY-SIZED PV SYSTEMS

As well as being fitted to the roofs of domestic houses, PV arrays can also be mounted on or integrated into the roofs and walls of commercial, industrial, school, community and public sector buildings. The additional capital costs of the PV system can, partially at least, be offset by savings in roof and cladding material that would otherwise have been incurred. Commercial, industrial, school and public sector buildings are mostly occupied during the daylight hours, which is when the highest levels of solar radiation are available and when electricity demand is highest. The electricity generated by a PV system will offset the electricity supplied from the local energy company, thereby reducing costs. A working demonstration of this is at Northumbria University in Newcastle, which installed a $40kW_p$ system on the front of a refurbished computer centre building in 1995, making it Britain's first building with PV cladding.

The output from PV systems during daylight hours can be highly variable, changing dramatically over short timescales with variations in cloud cover. The output over a longer period, however, is largely predictable and this would be of benefit to electricity grid operators. In Europe there are growing numbers of utility-sized grid-connected PV installations. One of the most significant is the RWE Kobern-Gondorf array on the banks of the Mosel River, close to the city of Koblenz. This installation has been a testing centre for a range of manufacturers' products from Japan, America, France and Germany. The array is $340kW_p$ and generates about 250,000kWh a year. As well as testing the output and electrical characteristics of panels, the installation demonstrates how the array integrates into the surrounding environment and has permitted the German utility to gain good large-scale PV operating experience.

A consortium including RWE has constructed a 1MW PV plant near Toledo in Spain. High-efficiency BP Solar, monocrystalline panels make up half the array and the plant is operated in association with a hydroelectric scheme. A large-scale 100MW utility-sized installation has been proposed for a former nuclear test site in the Nevada desert.

PV MARKET DEVELOPMENT

The PV market is very much focused on reducing the costs and increasing production, to give better PV availability. According to the *European Union PV Status Report* in 2006, global production is very much led by demand, with growth rates of 45 per cent per annum. Demand for PV modules is highest in Germany, followed by California, Spain, Italy and Japan, giving a world total installed estimate of $1{,}759MW_p$. Current estimates put the market value at around 9 billion Euros for an industry employing around 70,000 jobs worldwide. The report predicts that the market will continue to grow at around 40 per cent per annum, estimating that it will be worth around 40 billion Euros by 2010. On the production side, this huge growth in demand has triggered a global shortage of silicon feedstock, which has presented an opportunity for the accelerated introduction of advanced production techniques, such as thin film PV and solar concentrator technologies.

Figure 76: Solarcentury's C21e range integrates fully and blends aesthetically into the roof of the Owenvale Court BIH Housing Association in Belfast. The installation has 19kW$_p$ of C21e panels. (Author/ www.solarcentury.co.uk)

A similar analysis carried out in 2005 for the PV market in the UK is the IEA/DTI *National Survey Report of PV Power Applications in the UK*. This identified that 2,713kW of PV had been installed in 2005, which was a 33 per cent increase on 2004, bringing the total installed UK capacity to 10.9MW$_p$. For larger projects the average prices paid were in the range 4.5 to 7 Euros per kW$_p$, and for smaller grid-integrated schemes, in the range 1kW to 3kW, the price was around 8.5 Euros per kW$_p$. PV Crystalox Solar, which manufactures multi-crystalline blocks in the UK, saw production rising by 32 per cent to 250MW in 2005. The UK's only indigenous cell-manufacturing facility, ICP Global Technologies at Bridgend in South Wales, increased production from 1.5MW to 1.8MW in 2005. Sharp employs 250 people to assemble crystalline silicon modules at Wrexham in Wales and they produced 39.8MW in 2005. Specialist glass manufacturer, Romag, produced 0.5MW of semi-transparent crystalline PV laminates during 2005 at its facility in Consett, County Durham. Much of the output from these plants went to the global export market.

FUTURE TECHNICAL DEVELOPMENTS

In terms of advances in solar cells, current research falls into three main areas:

- The reduction of the cost of the cells and their manufacture, at the same time increasing cell efficiency, to make them more competitive with other energy sources.
- The development of new technologies to provide new solar cell architectural designs.
- The development of new materials that can serve as light absorbers and charge carriers.

A principal research line, which will obviously benefit the silicon industry as a whole, is the development of manufacturing techniques that will produce very pure silicon. In the late 1990s a process was developed that reduces solid silica to pure silica by electrolysis and this may offer new opportunities for cell development. The invention of conducting polymers and semiconducting plastics may see solar panels made from plastics in the future. Their efficiency is currently much lower, at around 6.5 per cent, but

Table 33: PV installation golden rules.	
Top Tips for Installing PV	**Things to Avoid when Installing PV**
Make sure the PV panels are orientated as closely as possible to south and tilted at an angle of approximately 10 degrees less than the latitude angle.	Do not install PV panels on a roof that faces directly west, east or north. Panels can be mounted on the ground or on flat roofs using suitably orientated and tilted mounting frames.
Contact the local electricity supply company to establish the options for selling the spill electricity and receiving the green value for electricity generated (ROCs and LECs), so as to maximize the revenue.	PV panels normally require a Building Control inspection and Certificate – so don't forget to contact the local Building Control office before work is commenced.
If the building is listed or in a conservation area, then early contact with the planning authority is recommended.	Do not place PV panels where they can be overshadowed by trees, buildings or other structures – allow for shadow-fall throughout the day and across the seasons.
If a beneficial spill tariff is not available, use timers or operate non-time dependent electrical appliances (dishwashers, washing machines, irons, etc.) when there is an output from the panels to maximize savings against electricity demand.	Do not forget the output from PV panels can change rapidly over the space of a few minutes, since a cloud passing suddenly across the Sun can result in a rapid drop.

the hope is that these cells can be mass-produced in a low-cost process. However, all organic solar cells produced to date have suffered from rapid performance degradation through exposure to UV light and have unacceptably short lifetimes.

Most conductive polymers discovered to date are very sensitive to atmospheric moisture and oxidation, making commercialization impracticable at the moment. At the cutting edge of these researches, carbon nanotubes or nanocrystals are showing some potential in the form of quantum dot modified photovoltaics, which can achieve very high efficiencies. Carbon nanotube materials are flexible and can be deposited on surfaces in a variety of ways.

If the large-scale deployment of PV is to be considered, then cost is the most important consideration. Currently it is prohibitive for anything but demonstration schemes or installations supported through grant initiatives. The land area required for very large arrays could become a significant factor. From an electricity grid operators' perspective the maximum contribution from PV systems will be made during the summer months when demand for electricity is lowest and the contribution is least in winter when the demand is highest. Another issue for PV systems is understanding how best to integrate small- to medium-scale systems into the electricity distribution system as embedded generation sources.

Solar PV can deliver clean, silent electricity, and this makes it very suitable for urban use, particularly given the UK government's objective of zero carbon housing by 2017. Despite the fact that it has the potential to meet a significant proportion of future electricity needs, its current high relative cost means that it is only likely to make a small contribution to the UK's 2015 renewables target.

PV SYSTEMS GLOSSARY

Anode A terminal in a battery or electrical circuit to which the electrons are attracted.

Arsenic (As) A semi-metallic element. Arsenides are compounds of arsenic.

Bulk material/bulk silicon Bulk is the term used to indicate that there is sufficient material to create a number of solar cells.

Cadmium (Cd) A metallic element, frequently termed a heavy metal.

Carbon nanotubes A carbon nanotube is a one-atom thick sheet of graphite (called graphine) rolled up into a seamless cylinder with a diameter of the order of a nonometer (one thousand millionth of a metre). Carbon nanotubes exhibit extraordinary strength and have unique electrical and heat conduction properties.

Cathode A terminal in a battery or circuit to which positive charge carriers (e.g. positive holes) are attracted.

Charge carriers These are positive holes and electrons.

Connection agreement An agreement between the local energy supply company and the customer who installs the PV panels.

Conventional power sources Sources from which we have traditionally got our electricity (coal, oil and gas power stations).

Current collector Metal strips, visible in the PV panels, that collect and conduct the electricity generated away from the cells.

Diffuse Indirect solar radiation, sunshine through clouds.

Doping The introduction of other materials or elements to affect the electrical characteristics of silicon.

Electricity grid/grid system The electricity network, including transmission lines, transformers, lines, cables, switches and other associated equipment.

Electrolysis The application of electricity to a liquid or molten material to separate out the constituent components.

Engineering Recommendation G59 Recommendations for the connection of embedded generation plant to the regional electricity companies' distribution systems.

Engineering Recommendation G83/1 Recommendations for the Connection of Small Scale Embedded Generators (up to 16A per phase), in Parallel with Public Distribution Networks (issue 1:2003) – electricity company regulation covering small PV systems.

Gallium (Ga) A metallic element useful in semiconductor applications.

Grid connection cost Usually means the cost of connecting a system to the electricity network, including lines, cables, transformers, switches and protection equipment.

Heliostat A mirror that tracks the movement of the Sun to reflect its radiation onto an absorber.

Light absorption co-efficient A measure of how easily a material absorbs light.

Load cycle When the equipment is turned off or switched on, and how much electricity it uses when switched on.

Parallel connection A method of connecting solar panels with the input and output connections linked together.

Photon Represents a single bundle of light energy.

Positive hole Created in a semiconducting material when an electron is displaced or removed.

Quantum dots These offer substantial potential for new developments in semiconducting materials, such as silicon, and for next-generation PV technology.

Ribbon-form silicon Silicon drawn out from a molten crucible into a ribbon or long narrow strip.

Ruthenium (Ru) A metallic element used in fluorescent dyes.

Selenium (Se) A non-metal element similar to sulphur.

Series connection A method of connecting the output of one solar panel to the input of the next one.

Silicon (Si) A non-metal element, the principal component of sand. P-type and n-type silicon are types of silicon that allow for the creation of positive holes or electrons, respectively.

Solar radiation, Solar irradiance, insolation The radiation arriving at the Earth's surface from the Sun, usually expressed in kilowatts (kW).

Solar shingles A special type of solar PV panel integrated within fabric that resembles roofing tiles.

Solid state semiconductor Electronic devices or materials that operate as a consequence of the semiconducting properties of silicon.

Spill/exported electricity Electricity not used within the premises and returned (exported) to the electricity network. PV systems spill back onto the electricity network at times when the panels generate more electricity than the premises are using.

Substrate Material used as a foundation for the application of other substances, usually glass, plastic or perhaps metal, which has the PV material applied to it.

Tellurium (Te) An element useful in semiconductor applications.

Tempered glass Glass that has been treated to harden it.

Titanium (Ti) A metallic element used in light-absorbing dyes.

Utility-sized systems Large-scale power systems that serve a large number of customers.

Vacuum spluttering A process that deposits molten silicon on substrate as a thin film in vacuum conditions.

Wafer/wafer suspension Several layers of thin silicon layers stacked on top of each other.

USEFUL PV CONTACTS AND WEBSITES

Companies and Trade Organizations

BP Solar
www.bpsolar.co.uk
PO Box 191, Chertsey Road, Sunbury on Thames, Middlesex TW16 7XA
Tel: 01932 779543
UK division of a leading world and European energy company.

DayStar Technologies
www.daystartech.com/
US company producing a range of PV cells, including lightweight films.

European Photovoltaic Industry Association (EPIA)
www.epia.org/
Website of the association of European PV manufacturers.

General Electric Company
www.gepower.com/prod_serv/products/solar/en/index.htm
Information on the US-based company's PV manufacturing activities.

International Energy Agency (IEA) Photovoltaic Power Systems Programme
www.iea-pvps.org
Reports, statistics, photographs and other useful information on developments in photovoltaic power systems in IEA member countries.

Natural Resources Canada
www.retscreen.net/ang/home.php
Free software to assist in designing Solar PV (and other renewable) energy systems.

Photon International
www.photon-magazine.com
Website of the leading magazine on photovoltaics.

PV in Social Housing
www.itpower.co.uk/pvish/index.htm
EU-funded website providing details of PV installations in social housing in Austria, Germany, the Netherlands, Spain and the UK.

Renewable Energy Association (REA)
www.r-p-a.org.uk/home.fcm
Membership merged with the British Photovoltaic Association in 2006. The latter's online information service is available at:
www.greenenergy.org.uk/pvuk2/uk/index.html

Sharp Corporation
www.sharp.co.uk/page/solarproducts
The world's largest PV manufacturer. Production is mainly based in Japan, but the company also owns the UK's only (currently) PV manufacturing plant, in Wales.

Shell Solar
www.shell.com/solar
Shell Gas & Power's renewable energy site, specifically its photovoltaic energy activities.

Solarcentury
www.solarcentury.co.uk
UK company specializing in PV system installation.

Sustainable Energy Ireland (SEI)
www.sei.ie
Offers advice on PV systems in the Republic of Ireland.

US National Centre for Photovoltaics
www.nrel.gov/pv/
The leading US photovoltaics research centre, part of the National Renewable Energy Laboratory (NREL) and funded by the US Department of Energy. Links to research literature on PV.

SOLAR RECORDS AND DATA

Met Office
www.metoffice.gov.uk/

Climate data (including sunshine on a monthly basis) can be found at:
www.metoffice.gov.uk/climate/uk/stationdata/index.html
www.metoffice.gov.uk/climate/uk/averages/19712000/index.html

Records for England and Wales
Met Office, FitzRoy Road, Exeter, Devon EX1 3PB
Tel: 0870 900 0100 Fax: 0870 900 5050
Email: enquiries@metoffice.gov.uk

Records for Scotland
Met Office, Saughton House, Broomhouse Drive, Edinburgh EH11 3XQ, Scotland
Tel. 0131 528 7311

Records for Northern Ireland
Met Office, Cargo Complex, Belfast International Airport, County Antrim BT29 4AB, Northern Ireland
Tel: 028 9441 7050

Records for Republic of Ireland
Met Éireann Headquarters, Glasnevin Hill, Dublin 9, Ireland
Tel: 00 353 1 8064200 Fax: 00 353 1 80642
http://www.met.ie/contactus

SOLAR PV PUBLICATIONS AND BOOKS

British Photovoltaic Association, 'Photovoltaics in the UK: Facing the Challenge, A Strategy for the UK to Claim a Significant Share of this Dynamic Clean Energy Industry' (January 1999); available at www.greenenergy.org.uk/pvuk2/reference/strategy-textonly.pdf

Lysen, E., amd C. Daey Ouwens, 'Energy Effects of the Rapid Growth of PV Capacity and the Urgent Need to Reduce the Energy Payback Time' (2002); available at www.copernicus.uu.nl/uce-uu/downloads/Presentaties/OD84_LysenDO.pdf

Pieper, A., *The Easy Guide to Solar Electric: For Home Power Systems*, 2nd edn (Adi Solar, 2001) [US systems]

Strong, S.J., and W.G. Scheller, *The Solar Electric House: Energy for the Environmentally Responsive*, Energy-Independent Home, 3rd edn (Chelsea Green Publishing, 1993) [excellent beginner's guide]

Thomas, R., *Photovoltaics and Architecture* (Spon Press, 2001) [excellent primer on the subject of integrating photovoltaics into buildings in the UK]

CHAPTER 7

Hydroelectricity

If you are fortunate enough to live beside a river or a fast-flowing stream, a small-scale hydroelectricity installation may provide an opportunity to achieve significant savings on your electricity bill and reduce your carbon footprint for many years to come.

ELECTRICITY FROM RIVERS AND STREAMS

Hydroelectricity is the most widely used of the renewable technologies, with almost one-fifth of the world's electricity generated from water power. In some third-world countries, hydroelectricity provides around 30 per cent of their demand for electricity. The largest hydroelectricity generating plants are in countries with high mountains and fast-flowing rivers, such as Canada, United States and China. These large-scale schemes draw water from reservoirs restrained by dams, although it is also possible to generate electricity from the flow of water in rivers and streams. The Netherlands, for example, although a predominantly flat country, has around 40MW of hydroelectric generating stations. The largest hydroelectric station currently operating in the world is the Itaipu project, a 12,600MW joint scheme between Brazil and Paraguay. Water stored behind dams means that it is available for electricity generation at times of greatest need; however, it suffers the disadvantage that large areas of land, indeed sometimes several villages, are lost in the inundation and subsequent water storage. Pumped storage schemes have been developed to pump water into reservoirs at times when there may be a surplus of electricity available and then to release it to generate electricity when it has the greatest value.

Small-scale hydroelectricity is much less environmentally intrusive, generating electricity from reasonably fast-flowing rivers and from small streams with high heads but low flow. These small-scale run of river schemes are the main subject of this chapter. Most of the hydroelectric sites in the United Kingdom and Ireland with a potential of generating one megawatt of electricity and above have already been developed, but there are still a significant number of locations capable of producing electricity in the range 100kW to 500kW. With modern developments in micro and pico (very small) hydro technology, an even larger number of smaller sites in the range 10kW to 50kW are now considered economically viable.

THE HYDROLOGICAL CYCLE

Although we consider hydroelectricity to be a water-based source of renewable energy, ultimately the energy is a result of the Sun's heat. Clouds are formed by evaporation from the Earth's water resources – the oceans, rivers and lakes. These clouds are carried by the wind over land, where hills and mountains will force them to rise to higher, cooler levels. Once cooled, the clouds will start to condense back to water, falling as rain or snow. This rain runs off the hills, forming streams and rivers (watercourses), which eventually find their way back to the sea and so the process continues in the so-called 'hydrological cycle'. These streams and rivers, along with water channels and reservoir infrastructure, make up the resource for the small-scale hydro schemes under consideration in this section.

A BRIEF HISTORY OF HYDROPOWER AND HYDROELECTRICITY

Waterwheels have been in use for several thousand years, initially providing irrigation for land in the Middle East and China, but water has also been used traditionally as the motive power to drive corn and grain mills. By the eighteenth century waterwheels were widely used to drive the machinery of the Industrial Revolution, operating manufacturing equipment such as bellows, textile looms and forges. The power developed by these early machines was quite modest by today's standards, with outputs in the order of a paltry few watts to several tens of kilowatts. With the advent of water turbines in the mid-nineteenth century, the versatility of water-driven machinery increased significantly. Turbines were smaller, more compact, had greater efficiency and could run at high speeds. They were also much better suited to the requirements of electricity generation. Technological improvements have continued, allowing modern hydro turbines to be classified as small-scale turbines (with powers up to 5MW), micro-turbines (with ratings below 500kW) and even pico-turbines with outputs of tens of watts.

With these advances, useful power outputs can be obtained from fast-flowing water in small streams. Small turbines or waterwheels generating a few kilowatts may not be grid-connected but instead may be used to charge batteries. With sufficient capacity, batteries could provide self-sufficiency in electricity. Very small, grid-connected hydro schemes for domestic use are uncommon, although small generators of 25kW and above can be used in commercial applications.

Table 34: Turbine classification.

Turbine Electrical Output	Designation
5MW (UK), 10MW (Europe)	Small-scale
< 2MW	Mini
< 500kW	Micro
< 10kW	Pico

Although these classifications are purely arbitrary, they are a useful means of describing the output range. It is also worth noting that small turbine output is not reflected in physical dimensions, for example a 200kW turbine might have a turbine diameter of 2 metres.

HYDRO BASICS

Energy is produced from a hydroelectricity scheme when water from a stream, river or other watercourse is diverted through a turbine. The potential energy (PE) from the water's head or height and the kinetic energy (KE) from the water flow or movement in the stream are captured to turn a turbine and produce the electrical energy. The head is sometimes referred to as the fall. The rotating turbine is usually connected to a generator through a shaft and gearbox. The size of the electrical output from the generator is dependent on the energy carried by the water passing through the turbine and the efficiency of power conversion.

RIVER FLOW DURATION CURVE

One of the most important pieces of information that can be used to assess the possible energy output from a proposed hydro development is the flow duration curve for the watercourse. For a river, the curve will show how the flow changes throughout the year. Hydrologists can produce a flow duration curve for a river by studying the recorded water flows, preferably over a number of years. The curve is a graphical representation of the probability of the number of days in a year that a particular flow will be exceeded (expressed as the percentage exceedance probability). Energy (kWh) is a measure of the length of time over which power (kW) is produced. The energy output (electricity) from the hydro is key, since it is this

Figure 77: River head (potential energy) and flow (kinetic energy) carry the energy that is converted by the hydroelectricity turbine.

which will be used to bring in the revenue (based on the electricity generated). A site on a river with a variable flow and a high peak flow may not have the potential to produce as much electricity as a river with a higher mean flow but a lower peak flow.

COMPONENTS OF A SMALL-SCALE HYDRO SCHEME

A typical hydro scheme will have the following components:

- An intake – this is often associated with a weir, sluice or dam, with the intake receiving water from the watercourse above the weir, sometimes referred to as a lade (or layde).
- Headrace – a short open channel that conveys the water from the intake to the penstock. Long headrace channels are uncommon in modern schemes due to environmental and economic constraints.
- The penstock – the section of pipe that conveys the water to the turbine inlet. The pipe should be of a large diameter to minimize losses through friction.

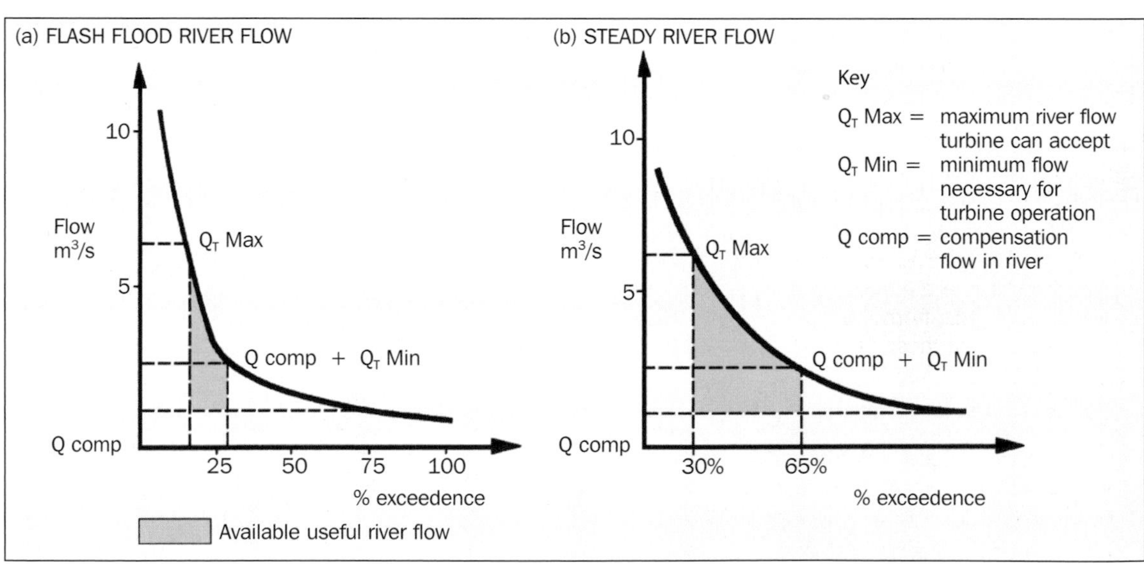

Figure 78: Flow duration curves of two rivers. A flash flood in river (a) produces a high flow rate over a short period of time. River (b) has a steady flow rate that produces a lesser, but more useful flow. The area below gives an indication of the energy potential of the river.

Output Power from Hydro Schemes

The amount of energy that is available from a hydro scheme is dependent on a number of factors:

- The available flow of water that can be directed through the turbine, usually represented as Q, the flow rate in m^3/s or cumecs (one cubic metre is 1,000 litres). For environmental and fish protection reasons, it is good practice not to take the full river flow through the turbine.
- The head is the height through which the water falls. The gross head is the difference between the upstream water level, where the water enters the turbine, and the downstream level, where the water exits the installation. The net head (H, in m) is the head across the turbine only and is the gross head less the losses (through friction) in the inlet channel, associated pipework, and the losses through the screens.
- The force of gravity, which is a constant (g, in m/s^2).

Modern, well-designed hydroelectric installations can be very efficient in converting the power in the water to an electrical output; efficiencies in excess of 85 per cent are possible, although with older machines it will more likely be closer to 50 per cent. The electrical power output for a given site is the product of the head, the design flow rate and the efficiency, and can be estimated using the formula below. The energy available for a given site (in kWh) is the product of the power produced and the time over which this is available, which can be obtained from the river's flow duration curve (see page 160). The head for most sites remains constant but the flow rate is highly variable. It is possible to take a number of average flows, over short periods, from the flow duration curve and the energy at these flows can be calculated (by multiplying the flows by the time they are available). Specific turbine designs will work more effectively over a restricted range of flows and it is important to take this into account when selecting the turbine. With very small schemes, it may be convenient to measure the flow in litres per second; if this measure is used the resulting power will be in watts (W), rather than killowatts (kW).

The output power from a turbine is given by the expression:

$$P = \rho\, g\, H\, Q\, \eta_t$$

where P = turbine shaft power or rated output (kW), ρ = water density (kg/m^3), g = acceleration due to gravity (m/s^2), H = net head (m), Q = flow (m^3/s), η_t = turbine efficiency. The electrical power output from a turbine and generator is given as:

$$P_e = P\, \eta_g\, (\eta_{gb})$$

Where P_e = electrical output (kW), P = turbine shaft power (kW), η_g = generator efficiency (%), η_{gb} = gearbox or speed increaser efficiency (%).

A general approximation may be assumed for the value of the constants (ρ, g, η_t) in the equation above. A value of 6 is indicated, although different sources will quote this approximation as a value from 5 to 10. Accurate flow data and careful analysis of the flow duration curve, pipeline and channel head losses can give a more accurate figure for the annual energy output, but, as a first approximation, the equation below can be applied.

$$P_e = 6\, H\, Q$$

An example for a medium to high head site, where the gross head is 110m, with head losses of 10m, means that the net head is 100m; with a flow of $1m^3/s$, this will give an approximate turbine output of:

$$P_e = 6 \times 100 \times 1 = 600 \text{ kW}$$

For a medium to low head site, with a net head of 11m, assuming losses of 1m, and where the flow is $10m^3/s$, then:

$$P_e = 6 \text{ x } 10 \text{ x } 10 = 600 \text{ kW}$$

So in these simplified examples, despite the lower head, the higher flow will give the same electrical output. As a first estimate, multiplying this number by the numxber of hours in a year (8,760) times the anticipated load factor – say, 40 per cent (for an explanation of load factor, see Chapter 1) – will give a rough estimate of the likely energy output from the system in kWh. In this example the approximate output would be around 2,102MWh.

- The turbine house – houses the turbine, gearbox, electrical switching, electrical control devices (PLC) and protection relays, control panels, power factor correction equipment and generator.
- The outflow – through which the water is exhausted back to the river.
- Electrical connection – the electricity generated must be fed to the grid or to the associated premises, through an overhead line or underground cable.

The size and extent of the components identified above will depend entirely upon the conditions of the site under consideration. The intake is typically taken from above a concrete or rubble masonry weir stretching across the river, and perhaps up to 2 metres high. A bypass or spillway will be incorporated, which will ensure that the watercourse is never totally deprived of river flow (the remnant flow is known as the compensation flow – shown as Q comp in Figure 78), and a screen or trashrack prevents floating debris or large fish from entering the penstock. A sluicegate or control valve is often incorporated at the intake. Where there is the possibility of a high silt load in the river, a settling tank may be required to encourage the silt to settle out. The penstock is a pipe of diameter between 10cm and 100cm, typically made from steel, plastic or composite material. The penstock length

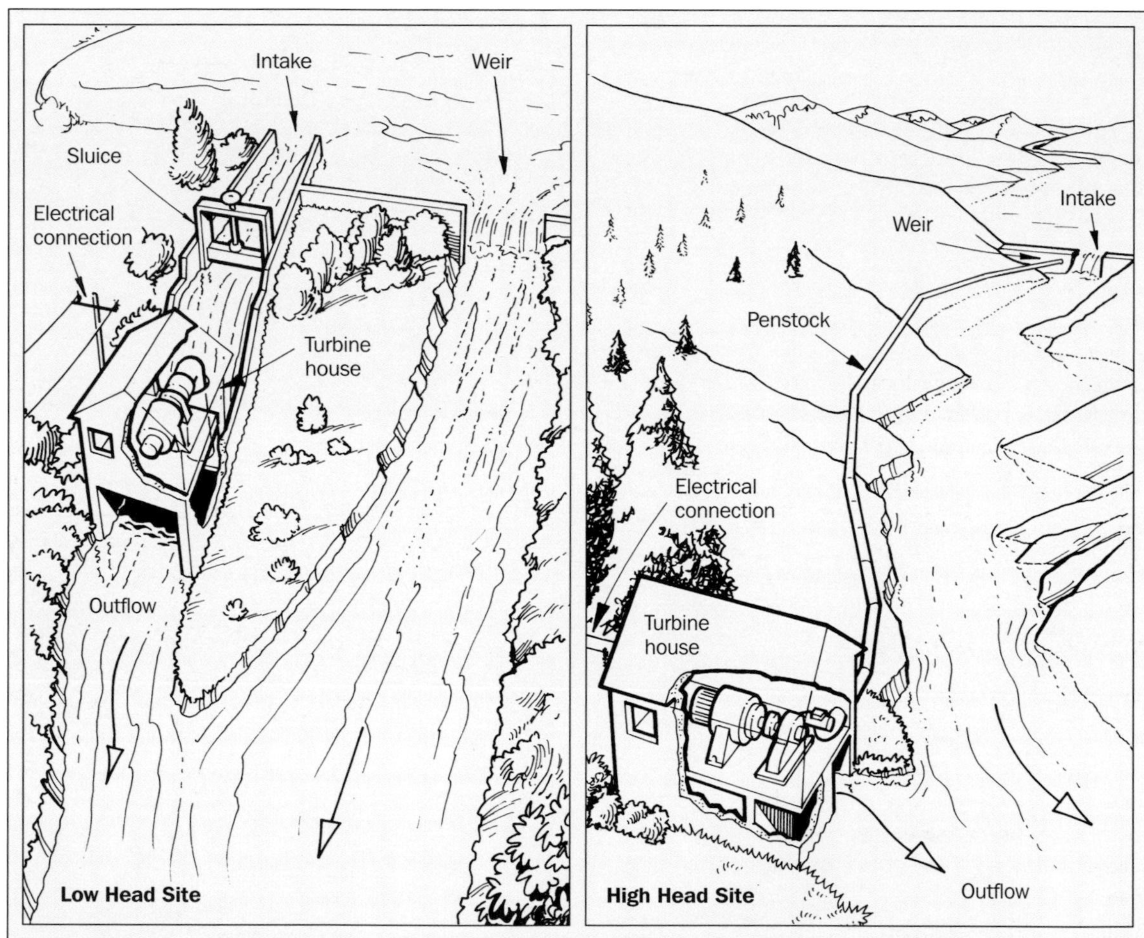

Figure 79: Small-scale hydroelectricity system components. (Illustration reproduced and modified courtesy of Newmills Hydro Turbines)

Table 35: Appropriate turbine designs for selected head heights.		
Classification	**Head**	**Turbine Design**
High head	Above 100m	Pelton, Turgo, high-head Francis
Medium head	20m to 100m	Francis, cross-flow
Medium to low head	Less than 20m, but above 5m	Francis, cross-flow, propeller, Kaplan
Low head	Below 5m, but above 3m	Cross-flow, propeller, Kaplan
Ultra-low head	Below 3m	Propeller, Kaplan, waterwheel
Note: Turbines can be engineered to perform at varying water flows with only a small loss in efficiency. In most applications, turbines are specifically manufactured for a given site.		

and diameter will depend very much on the nature of the scheme. Run of river schemes tend to have wide, short penstocks. High head schemes typically have long, smaller diameter penstocks, perhaps in excess of a kilometre in length. Long penstock pipe runs are normally buried underground and restraining anchor blocks are used where vertical and horizontal changes in direction occur.

Open headrace channels are common on refurbishment schemes based on old watermill sites, but they are rare on new projects. Turbine houses are generally unobtrusive, typically single-storey buildings resembling a small cottage located close to the riverbank (to minimize tailrace length). The water is returned to the river through a concrete channel or masonry tailrace between the turbine house and the watercourse. To avoid back-flooding of the turbine, the tailrace should have a gradient sufficient to encourage the free discharge of water into the river.

WATER TURBINES

Hydro turbines typically consist of a wheel or runner, which is connected to a shaft. When the water acts on the runner, the shaft spins in bearings and operates through the gearbox to drive the generator. The rotational energy of the shaft is converted to electrical energy. A range of water turbines has been developed over the years to accommodate the variety of flow and head conditions encountered. They may be classified into two categories, impulse turbines and reaction turbines, and in subgroups based on available head, as in Table 35.

REACTION TURBINES

In reaction turbines, the water flows across the blades or propellers and there is no water jet. The flow of water over the blades causes a pressure difference across them, producing a spin force (or reaction force) that rotates the shaft. Reaction turbine runners resemble ships' propellers.

IMPULSE TURBINES

Impulse turbines are designed to capture the energy from a high-speed jet of water using specially shaped cups. The water strikes the cups (receiving an impulse) and they can capture almost all of the energy in the water. The water then leaves the turbine to return back to the river. In some designs a number of jets can be used to increase the effectiveness of the turbine under changing load conditions.

SPECIFIC SPEED

The Specific Speed is the principal parameter for matching a specific hydro site with the correct turbine type. Specific Speed (N_s or n_s) defines the turbine's shape in a way that is not related to its size. This means that a new turbine can be designed by scaling from another existing turbine of known design and performance. Turbine manufacturers always quote Specific Speed as part of the specification for their turbines and they will refer to the point of maximum efficiency. This allows accurate calculations to be made of the turbine's performance for a range of

Table 36: Specific speed related to head and turbine type.					
Single-jet Pelton	**Multi-jet Pelton or Turgo or Cross-flow**	**Francis**	**Kaplan & Propeller**	**Specific Speed (n_s)**	**Application**
■				7	High head
■	■			27	High head
	■	■		39	Medium head
	■	■		60	Medium head
		■		175	Medium head
			■	300	Low head
			■	900	Low head

heads and flows. Specific Speed is most useful when attempting to complete the design of a new turbine, since it can be used to calculate the physical dimensions of the component parts. Specific Speed (n_s) is given by the following expression:

$$n_s = n\sqrt{(P/H^{1.25})}$$

where n = rpm, P = the turbine shaft power (kW), H = the net head (m).

Another parameter that may be defined in the manufacturer's specification is the runaway speed of a water turbine, which is its speed at full flow with no shaft load attached. Turbines are designed to survive the mechanical forces experienced at this speed. Typical runaway speeds are shown in Table 37.

Table 37: Typical runaway speeds.	
Turbine	**Runaway Speed n = Turbine Speed in rpm)**
Pelton turbines	1.8 × n
Francis turbines	1.8 × n
Cross-flow turbines	2.0 × n
Kaplan turbines	2.6 × n

TYPES OF HYDRO TURBINE

Francis Turbine

Francis turbines are the most common water turbines in use today. They operate in schemes with heads within a range of ten metres to several hundred metres and are primarily used for electrical power production. Francis turbines can be mounted both vertically and horizontally.

The Francis turbine is a reaction turbine, which means that the water pressure changes as it moves through the turbine, giving up its energy. The turbine is situated between the high-pressure water source and the low-pressure water exit. The water inlet is spiral-shaped and guide vanes direct the water tangentially onto the runner. This radial flow exerts a force on the runner blades, causing the runner to spin. The guide vanes (sometimes called the wicket gate) may be adjustable to allow for efficient turbine operation over a range of water flow conditions. Francis turbines do not perform well under 40 per cent of design flow with efficiency dropping off rapidly with reducing flow.

Kaplan and Propeller Turbines

Kaplan turbines are currently the most widely used turbine throughout the world for high-flow, low-head applications. The Kaplan turbine is a propeller-type turbine with adjustable blades. It was developed from the Francis turbine in 1913 by an Austrian professor,

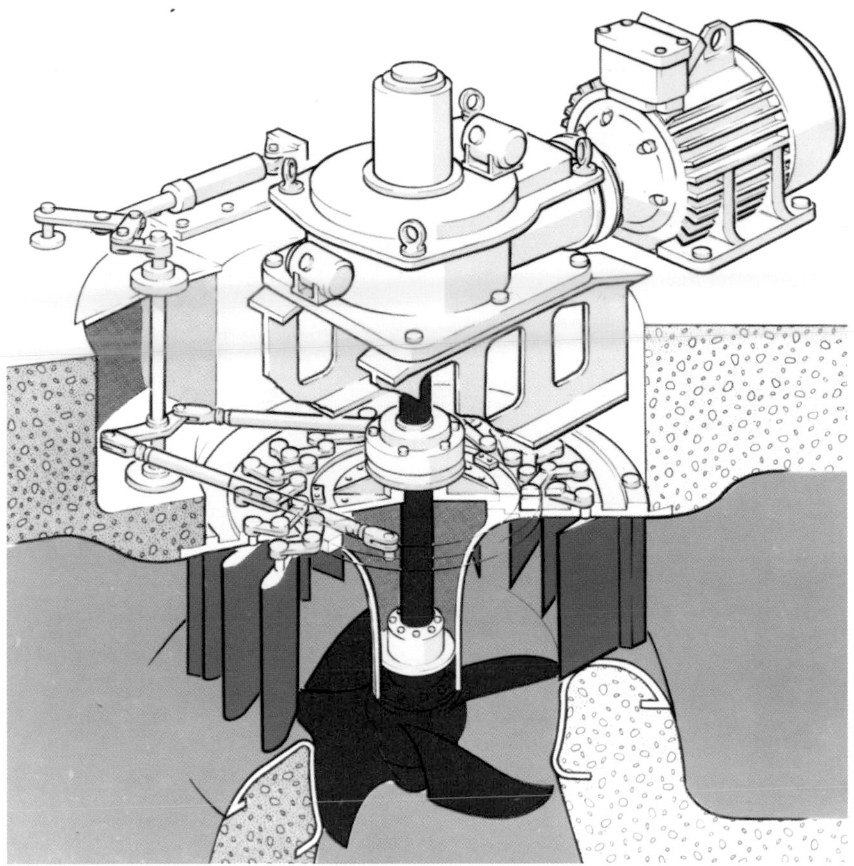

Figure 80: Cutaway diagram of a vertical Kaplan (propeller) turbine showing gearbox and generator. (Illustration courtesy Newmills Hydro Turbines)

Viktor Kaplan. Kaplan turbines permit efficient power production in low-head applications that are not possible with Francis turbines.

The Kaplan turbine is also a reaction turbine with an inward flow, which means that the water changes pressure as it moves through the turbine giving up its energy. The inlet to the Kaplan turbine is a scroll-shaped pipe, known as the spiral casing, and this wraps around the turbine's guide vanes. The spiral casing channels the water through the guide vanes onto a propeller-shaped runner, causing it to rotate. The outlet pipe is a specially designed draft tube, which slows down the water as it leaves to capture its kinetic energy. The turbine does not need to be at the lowest point of water flow as long as the draft tube remains full of water.

The variable geometry of the guide vanes allows Kaplan turbines to operate over a range of flow conditions. They are widely used at the lowest head hydro sites and are best suited for producing electricity in high-flow conditions. These applications are further developed below. Inexpensive micro turbines are available for small-scale power production with as little as one metre of head. Large Kaplan turbines operate at the highest possible efficiency (typically over 90 per cent) for decades, but they are more expensive to design, manufacture and install than other designs.

A number of variations on the basic Kaplan design exist, including:

- Propeller turbines, which have non-adjustable propeller blades. They are used in low-cost installations with small outputs. These turbines are widely installed and can produce several hundred watts from a head of only one metre.
- Bulb or tubular turbines sit snugly in the water delivery tube or inlet channel. A large bulb, which

Figure 81: Final factory assembly of a 1,000kW, four-jet Pelton turbine for Garbhaig Hydro, Loch Maree, Rosshire, Scotland. Image shows the finely polished Pelton wheel and nozzles. (Photograph courtesy Newmills Hydro Turbines)

holds the generator, guide vanes and runner, is centred in the water passage.

- Pit turbines are bulb turbines fitted with a gearbox. This allows for a smaller generator and bulb.
- Straflo turbines have the water flowing along the turbine shaft in an axial direction. The generator is located outside the water channel and is connected to the turbine runner.
- S-turbines eliminate the need for a bulb housing by placing the generator outside the water channel. This is accomplished with an angled shaft that connects the runner and generator.
- Tyson turbines are fixed propeller turbines that can be immersed in a fast-flowing river, either anchored permanently to the river-bed or attached to a boat or barge.

Pelton Turbines

A Pelton turbine, or Pelton wheel, is one of the most efficient types of water turbine. It was invented by Lester Allan Pelton in the 1870s and is an impulse machine extracting the energy from a jet of water.

In a Pelton turbine, jet nozzles direct powerful streams of water against a series of spoon-shaped buckets, mounted radially, along the edge of a wheel. The buckets absorb the energy in the water from the jet and this imparts an impulse that spins the turbine. To ensure the forces on the turbine wheel are balanced, the buckets are mounted in pairs and this helps to deliver smooth, efficient energy transfer from the water jet to the wheel. Pelton turbines are made in a range of sizes and they are the preferred option for high-head, low-flow sites. Large-scale hydroelectricity plants use Pelton wheels that weigh many tons and are very efficient. The largest Pelton wheels are around 200 megawatts; at the other end of the scale the smallest Pelton wheels are only a few inches in diameter. Small Pelton turbines can tap power from mountain streams with 30 metres or more of head but only a few tens of litres per minute of flow. Pelton wheels can operate with heads as small as 15 metres and as high as 1,800 metres.

Turgo Turbines

The Turgo turbine was designed as an impulse turbine, suited to medium-head applications. It was developed in 1919 by the firm of Gilbert Gilkes & Gordon Ltd as a modification of the Pelton turbine and in certain applications it has advantages over the Francis and Pelton designs. Turgo turbine runners are less expensive to make than Pelton wheels. Unlike Francis turbines, they do not need an airtight housing. There are a number of large Turgo installations, but they are also popular for small-scale hydro, where low cost is important. Turgo turbines can have efficiencies up to 86 per cent. The runner from a Turgo turbine looks like a Pelton wheel split in half and it is half the diameter of a Pelton runner for the same power (so twice the specific speed).

Cross-Flow Turbines

In the 1920s, several hydro system designers, including the Australian Anthony Michell, the Hungarian engineer Donát Bánki and the German Fritz Ossberger, combined to produced what has come to be known as the cross-flow turbine. It consists of a cylindrical water wheel or runner with a horizontal shaft. The runner has a series of semi-cylindrical, sharpened blades arranged radially. Some turbines have as many as thirty-seven blades and each blade is held in place by two end plates. The semi-cylindrical blades are welded to support plates at the blade ends. The water jet is directed onto the cylindrical runner at an angle of around 45 degrees by a fixed nozzle, and this transfers the water's energy to the blades. The turbine is regulated using the guide vanes to change the cross-sectional area of the flow passing through the turbine. The two profiled guide vanes divide and direct the flow in such a way that it enters the runner smoothly, regardless of the width of the opening.

DESIGNING LOW HEAD HYDRO SCHEMES

Low Head, Run of River Sites

Many low head hydro schemes involve the restoration of an existing watermill site, where most likely there will be an existing weir or sluice, and in many cases there may also be concrete for the turbine support and water flows. Assuming the condition of these items is reasonable, then savings can be made on the civil engineering work; however, it can also be restrictive on the water flow volume for a new scheme. Civil costs are often a deciding factor when considering the viability of a scheme. In low head projects, a high volume flow is required to achieve a reasonable energy output. Large flows require large flow channels, with associated high civil engineering costs. The turbine may be large and need to be accommodated within a restricted vertical height. Siphon turbines (discussed below) offer a solution to this particular scenario. Another factor to consider in low head, run of river schemes is that when river flow is high, the tailwater level can rise, reducing the net head. A reduction in head leads to a lower flow through the turbine, resulting in a loss of energy output.

Figure 82: This small-scale hydro turbine house contains the necessary control panels, valves, instruments, indicator panels and controls. It is designed to blend into the environment.

In high river flow conditions, trash and debris can build up quickly and consequently its removal becomes important. A trash rack and trash rack cleaner need to be sturdy enough to stop large items, such as floating trees and branches, entering the turbine, but the spacing between the grating bars needs to be wide enough to ensure flow resistance is kept to a minimum. Small items, such as leaves and sticks, can be allowed to pass through the turbine, but every precaution should be taken to prevent fish from entering the intake.

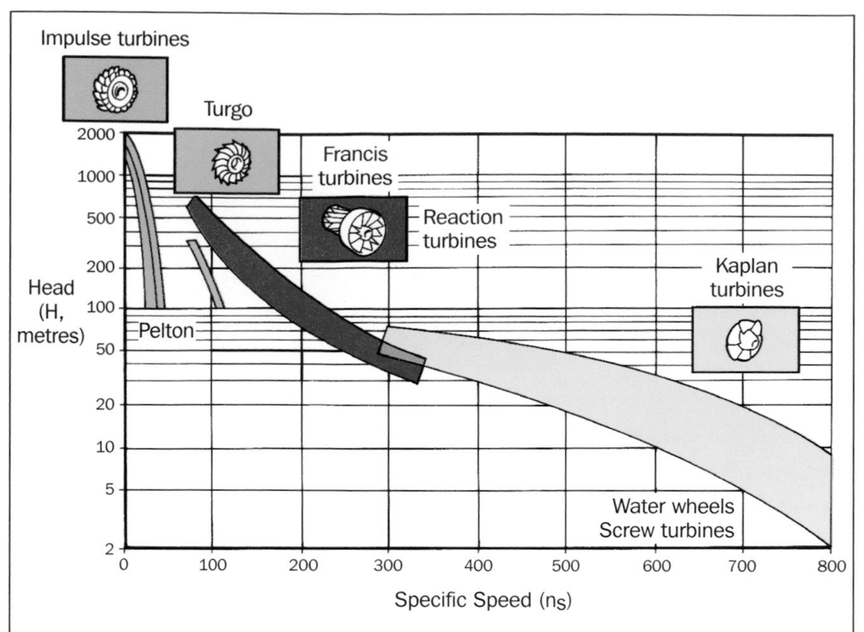

Figure 83: Turbine selection parameters: micro and pico turbines will operate at low to zero head.

Selecting Low Head Turbines

The decision as to which turbine option is the most appropriate depends upon the parameters, head and Specific Speed. For high and medium head hydro sites the technology is well proven; however, selecting the best technology to exploit lower head sites (<5m head) is less well established. In many instances the Kaplan turbine and the variations described above can provide a solution, and these are used extensively in projects greater than 1MW. They are, however, less suited to smaller, low head projects. Remote, automatic operation and fine control of these turbines is carried out using a programmable logic controller (PLC).

Pico and Micro-Hydro Options

Variations on the designs detailed above have been derived to reduce installation and operational costs at low or ultra-low head sites.

Tubular Turbines

These use a steel tube to encase the propeller runner. The tube has a right-angled bend so that the shaft of the runner can be brought out to mate-up with the speed-increaser (gearbox) and generator. The geometry of the tube can be arranged to suit individual site conditions.

Siphon Turbines

In ultra-low head applications a special variant of the tube turbine, called a siphon turbine, may be considered. Siphon turbines have a number of advantages, including reduced excavation, cheaper and simpler control of shutdown (by breaking the siphon), straightforward fabrication and installation, as well as easier maintenance and inspection. The arrangement permits both the turbine and generator to be above the upstream water level, facilitating inspection and maintenance. Siphon turbines can, however, be noisier and more visually intrusive than other designs.

Open-Flume Propeller Turbines

The open-flume turbine has no tapering intake pipework section to feed the water flow into the turbine; instead the guide vane assembly sits in a large open chamber with a spiral volute (casing) around the turbine, which can be made from concrete, brick or rolled steel plates. This design is very suitable for replacing old open-flume Francis machines in existing mill structures. Open-flume turbines have fewer components, but considerable site work is required to assemble and construct the machine.

Pit-Axial Flow (Right-Angle Drive)

This turbine variant was devised as a low-cost alternative to the bulb turbine. The runner shaft passes into a sealed pit, which runs from the base of the intake up into the powerhouse. Water flow passes either side of the pit to reach the guide vanes and runner. The pit itself contains a right-angle gearbox, from which a vertical shaft rises into the powerhouse to drive the generator. Some installations replace the gearbox with a belt drive to reduce costs. This is rapidly becoming the turbine of choice where low-cost, low-head applications are being considered.

Submersible Turbines (Mini-Bulb Turbines)

In very small schemes, where the dimensions of the bulb become smaller, access becomes difficult for inspection and maintenance of the gearbox and generator. The solution has been mini-bulb turbines in which the generator is submerged in a small watertight bulb, in a similar arrangement to that used with a submersible pump. With submersible turbines, there is no requirement for a powerhouse above the turbine. Consequently, the generator is automatically water-cooled and the visual and noise impacts of the scheme are minimized.

Waterwheels

Waterwheels are well proven and are particularly suited to ultra-low head applications. They are not as efficient as turbines, although they are still a viable proposition for producing electricity. Waterwheels are simple to control, have an attractive appearance and are straightforward to construct. The major disadvantage is that they run relatively slowly and require a high ratio gearbox, or other means of increasing the speed, if they are to drive a generator. For low powers (less than 5kW) and for heads below 3m, they are worth considering.

HYDRO TURBINE MAINTENANCE

Hydro turbines are designed to run for decades, typically thirty to fifty years, requiring very little maintenance of the main components. Major overhauls are, however, necessary at intervals of several years. A maintenance schedule for the runners and parts in

Figure 84: Final polish of a 150kW, 400mm diameter Francis turbine runner. (Photograph courtesy Newmills Hydro Turbines)

Figure 85: 240kW S-type axial turbine located at Askeaton, Co. Limerick, Ireland. (Photograph courtesy Newmills Hydro Turbines)

contact with water would include removal, inspection and repair of worn components. Turbine blades and runners suffer from pitting as a result of cavitation, fatigue cracking and abrasion from suspended solids in the water. Damaged areas and steel components are repaired by welding, usually with stainless steel rods. Other elements requiring inspection and repair include bearings, packing boxes, shaft sleeves, servomotors, cooling systems for the bearings, generator coils, seal rings and the wicket gate (guide vane) linkages.

Electrical connections, power factor correction capacitor banks and circuit breakers require a minimal amount of maintenance, usually no more than occasional contact cleaning and inspection.

THE MARKET FOR SMALL-SCALE HYDRO

Hydroelectricity has been around for many years and it is considered to be fully commercialized, although there have been capital grants available in the UK for small-scale installations. Historically, hydro has been developed at a huge number of rural sites and many of these have been restored to produce electricity. There are still a large number of opportunities and the British Hydropower Association has identified canal systems and locks as possible new opportunities. The installation cost of small-scale hydro is heavily loaded up front, with the civil works and capital purchase of the equipment. This means that after the equipment is bought, the initial outlay is offset by low running costs and free fuel.

ASSESSING OPPORTUNITIES AND SITE INSPECTIONS

It is essential to make an early site inspection, and the site should be reviewed to assess the following information:

- The gross head between the upstream and downstream water level.
- Site access and proximity to roads, particularly access for machinery and light cranes. The presence of an existing mill does not necessarily mean that the site is suitable for a more modern installation, since the upstream water management may have changed over the years.
- An environmental assessment to anticipate problems, such as visual impact, noise, site flooding, possible erosion and potential siltation (the degree to which silt may become a problem).
- Assess the proportion of the river/stream flow that will be available for the turbine. Wide shallow rivers make it difficult to channel river flow through the turbine.
- Select a suitable location for the turbine house, if required.
- Check out electricity connection – how close and accessible is the nearest point of electrical connection. Look for an overhead line or underground cable route.
- Assess other possible water users and review the likely impacts of the proposed development on them.
- Identify landowners on both sides of the river.
- Identify fisheries' interests.
- Review existing river flow – remember that modern hydroelectric plants require reasonably fast-flowing water to turn the turbine at sufficient speed to generate electricity.
- On redundant sites, assess the existing civil works, penstock and mill lades.

River flow duration curves and other hydrology information for the UK can be obtained for large streams and rivers from the Environment Agency, or from the Institute of Hydrology at Wallingford, Oxfordshire. They maintain up-to-date web-based data. In the Republic of Ireland this information is held by the Office of Public Works.

Assessing how individual and community electricity requirements can be met from hydroelectricity can be addressed using a web-based sequential decision making (SDM) approach. The approach considers the barriers to the installation and links information through an economic assessment, which identifies a range of turbine options. The SDM iterative model assesses the options' suitability, based on the location and demand, and combines the different types of information in a way that supports decision making. The SDM system is structured into five components. The hydrological resource is modelled

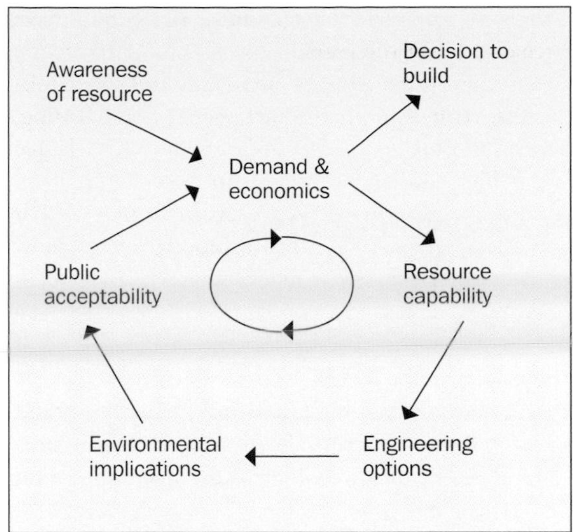

Figure 86: Work packages have been developed to enable the SDM model to be applied for hydro schemes. (Reproduced courtesy Dr Phil Leigh, Crichton Carbon Centre)

using Low Flows 2000, a specially developed software, and the turbine options are identified from hydrological, environmental and demand requirements. The consequences of different solutions are fed into other components of the model, so that the environmental impacts and public acceptability can be assessed and valued. (For further details on the SDM approach, see www.carboncentre.org.)

ECONOMICS OF SMALL-SCALE HYDRO

Costs are generally allocated into four main areas:

- **Plant and Equipment:** Includes the turbine, gearbox, belt drives, generator, trash rack, fish protection equipment and the water inlet control valve. Plant costs will be dependent on the site and the output from the scheme. High head machines have to pass less water than low head machines for the same power output. They also run faster and thus can be connected to the generator without the complication of gearbox or belts. As a general rule, high head scheme costs will be lower than for low head schemes of the same output.
- **Civil Works:** This is a major cost and includes the inlet channel, penstock, the turbine house, turbine foundations and outlet return channel back to the river. Civil works are very site specific. On high head schemes the major cost will be in the pipeline, whereas on low head schemes the costs will be due to the intake channel and water inlet control arrangements.
- **Electrical Works**: The control panel, power factor correction equipment, turbine house wiring, electric fish barriers (if fitted) and turbine control system. Additional costs may include the electrical connection to the grid, electrical protection (G31 and G59 relays and panel), meter costs and transformer, if not included in plant and equipment costs.
- **Additional Costs**: Might include design, consultancy, environmental impact assessment (EIA) report, costs of obtaining water licences, consents for water discharge, planning application fees.

Financial viability will be determined by many factors, including the output from the hydro scheme. Connection costs to the grid are dependent on location; this will determine proximity to suitable connection points and the capacity of the existing connection if one exists. Table 38 shows, for guidance only, a range of typical costs for a small-scale scheme, as compiled by the British Hydropower Association. Generally costs should not

Table 38: Typical spread of costs for small-scale hydro development.

	Low Head Sites (£1,000)	High Head Sites (£1,000)
Plant and equipment	30–90	15–60
Civil works	10–40	20–40
Electrical equipment	10–20	10–20
Other costs	8–15	8–15
Totals	58–165	53–135

be underestimated and the higher cost in the range will most likely apply. The annual electricity output is critical to the economic success of the scheme. Details of the payments for electricity exports and for the environmental value of the electricity (ROCs and LECs) are discussed in Appendix I.

Small-Scale Hydro Grants

In England, Wales and Northern Ireland the uptake of small-scale hydroelectric systems for domestic, community and other buildings has been encouraged through the introduction of capital grants. Grants have also been available to community and heritage groups in Scotland to set up hydro schemes or to restore old waterwheel sites. No grant support for hydroelectric systems has been available in the Republic of Ireland to date. Due to the difficulty in estimating uptake in these schemes, the duration and level of available grant funding is subject to change at any time. (For contact details for these schemes, their accredited products and approved installers, *see* pages 25–27.)

PLANNING PERMISSION AND THE ENVIRONMENTAL IMPACT OF SMALL-SCALE HYDRO

Some large-scale hydro schemes have been considered environmentally disastrous, however small-scale and run of river schemes are less obtrusive. Careful attention to ecological management in the design can result in schemes that have very little impact. It is important that other potential users of

Figure 87: Workshop assembly of axial-flow propeller turbines. This arrangement permits good part-load operation. (Photograph courtesy Newmills Hydro Turbines)

the watercourse are not neglected and that a meaningful flow is allowed to remain in the river once the turbine has drawn off its requirement.

In a life-cycle analysis of the energy balance, hydropower is considered to provide around 200 times the energy consumed during construction and operation. This high-energy yield is due to the long operating lifetime of the system, which can run for 100 years with minimal maintenance. Once installed the system produces no atmospheric emissions and only minimal impacts occur during normal operation and eventual decommissioning. Planning authorities may request details of how debris collected at the trash racks is to be treated and disposal off-site may be required. Care should be taken that there are no spills to the water during oil changes and component greasing. Some advocates suggest that hydropower operation can improve the water quality through increased oxygenation. A properly installed operating system normally emits little or no noise.

A planning application should include maps, diagrams and drawings showing the location and design of the intake, pipeline and turbine house, tailrace, security fence and lighting for urban schemes together with photomontages of the intakes. The grid-connection works, including the transformer and transmission line connection, should be detailed on the Ordnance Survey map section. Planners will also require details of vehicle access and movements, along with landscaping provisions and site management measures during the construction phase. Environmental Impact Assessments (EIAs) are generally required for applications above 500kW, or if an Abstraction Licence application is made. Early contact with the planning authority is recommended.

The most significant perceived impact of hydropower installations is on fish stocks. Fish living in rivers are adapted to compensating for a flow regime that naturally favours downstream movement. During much of their life cycle, they remain in areas where they are largely protected from the flow and are able to compensate for any downstream drift. For example, they may live among in-stream vegetation, stony areas or in deep pools where the water velocity is minimal. At certain stages of their life cycle, however, it is essential that fish migrate within the river system in order to reproduce or move to new feeding areas. During these migrations, inadvertent entry of fish into turbine intakes or out-falls can occur.

Fish, particularly migratory species such as salmon, trout and eels, are valuable assets in rivers. They support recreational and commercial fisheries, which contribute significantly to the rural economy. Their presence also adds significantly to the aesthetic and environmental value of the countryside. Healthy fish stocks support a diverse range of other wildlife such as otters and herons. As a consequence, any proposal to install a hydro turbine on a river is likely to be controversial and attract a great deal of interest from people who own or lease the fishing rights.

It is essential that fish are protected from physical damage or entrapment at water abstraction sites, and the developer should ensure that appropriate protection measures are built into the design of the hydro from the outset. If inappropriate measures are put in place at hydro-abstraction points, or the management of the fishery protection measures is slack, considerable long-term damage to fish stocks can occur. For this reason, fishery law has been framed to fully protect fish at all times with requirements for: the placing of gratings or screens to prevent fish moving into the turbine; construction of fish passes on weirs and impoundment structures; and restrictions on the abstraction of water to preserve the ecology of the river. National law is further reinforced by European directives such as the Water Framework Directive and the Habitats Directive, which requirse proper protection measures to be put in place and maintained throughout the life of the turbine.

At the outset, prospective hydro developers should contact the local fishery officer for advice and be prepared to consult with any fishery owner or angling club with an interest in the river.

ABSTRACTION LICENCES

Abstraction Licences are required to remove (or abstract) water from a watercourse, reservoir or lake. In the UK, Abstraction Licences must be obtained from the Environment Agency to redirect water from a

Table 39: Commonly available fish pass designs.		
Type of Fish Pass	**Target Fish**	**Operation**
Overspill and notch	Salmon and trout	A pool is constructed at the base of the dam or weir with a notch above it. This allows the fish to leap from the pool over the weir at the notch.
Diagonal bulk (diagonal wall)	All	A diagonal wall across the weir provides a deeper flow of water at the base of the weir. Sometimes a notch is placed immediately adjacent to the bulk.
Pool and traverse	All	A series of traverse or cross walls and pools are arranged to allow fish to pass over the weir in a series of easy stages. It is essential that the pass is supplied with a flow of water at all times.
Denil fish pass	All	Consists of a channel fitted with a series of closely spaced baffles, set at an angle to the axis of the channel. Named after its Belgian designer.
Source: O'Neill, G. and Connor, S., *Inland Fisheries Advisory Handbook No.1, Protecting Fish: Guidelines for Water Abstraction* (DCAL (NI), 2003)		

river and pass it through a turbine. The application for the licence can be complicated and sometimes expensive. Detailed flow data is required, including site flow predictions based on Environment Agency gauging station data, backed up with spot-gauging data and statistical analysis. This has to be done for average and wet years. A range of hydrographs and flow duration curves can be drawn from the data to show the effect of the hydro system on the local ecosystem. Licence and planning applications will require an environmental statement (sometimes an EIA) and the production of the necessary legal paperwork in order to demonstrate the right to abstract the water.

FISH PROTECTION

As noted above, any hydroelectricity development that involves the removal of water from a watercourse will need to provide the highest possible fish protection measures. The objective is to stop fish entering the turbine at the water intake and to discourage them from attempting to swim back into the turbine from the tailrace. In the UK this is a legal requirement under the Fisheries Act.

Although the Fisheries Act (available online at www.hmso.gov.uk/legislation) establishes a clear responsibility on water abstractors to protect fish, exception permits can be issued to provide alternative measures at any given site. The Water Order 1999 also provides a form of control on abstractors through the issuing of discharge consents.

Unless an exemption has been sought, most watercourses that contain stocks of the species mentioned above will require fish passes to be fitted. The design of the fish pass must be approved and it is recommended that the relevant authority (see below) is contacted before construction is started. The Fisheries Act calls for the regular closure of sluice gates, usually at the weekend, although this provision is outdated and best practice will allow plant shutdowns at times that better suit both the abstractor and the fish. For example, a four-week closure during the salmon smolt migration season is much more beneficial to both operator and fish than a statutory twenty-four-hour closure each weekend.

Fish Passes

An installed fish pass can provide an effective bypass, although fish will often attempt to climb the face of a dam or weir. A properly designed fish pass, however, will facilitate the passage of fish in most flow conditions or when fish wish to move within the river system. There are various types of fish passes in common use (*see* Table 39).

Preventing Fish Entering the Hydro Intake

Gratings or fish screens must be provided where the water is abstracted and returned to the river in order to prevent fish from entering the hydro plant. A fish screen should be placed at the hydro inlet to prevent fish being drawn into the intake. The entire surface of the grating must be covered with a wire lattice during certain periods to prevent small fish from entering the intake.

Fish may be attracted to the intake when they are migrating downstream. A well-designed fish screen will meet the following criteria:

- A self-cleaning grating or grid should be in place at the water intake. Recent experiments using sound arrays and bubble screens to divert fish away from the hydro intake have shown that they can work effectively.
- The water speed at the intake must be lower than the swimming capability of the fish. Fish swimming speed depends on a number of variables, including species, water temperature, size of fish and water quality. It is recommended that the velocity of water through a screen should not exceed 30 to 40cm/s for salmonid fish, and 15 to 20cm/s for coarse fish species, which are less active.
- An easily detectable escape route or fish pass, which will let the fish move away from the intake, must be included.

Designing a Fish Pass

The following points should be borne in mind when designing a fish pass:

- It is essential that the fish can find the entrance to the pass. Work on the river bottom should be carried out downstream of the dam or weir to direct fish towards the fish pass entrance.
- An adequate supply of water should be available to the fish pass at all times. Fish will not be attracted to a fish pass if the water flow through it is less than that coming over the weir.
- The speed of the water should be less than the swimming capacity of the fish.
- During periods of river flooding, the pass should still be active.
- The fish pass should operate without manual intervention.
- A well-constructed fish pass should not become blocked by debris or sediment.
- Fish passes should be constructed in a manner that prevents access by poachers.
- Pools in the pass should be of sufficient size to prevent fish becoming disorientated and of sufficient height to prevent the fish leaping out.

Preventing Fish Entering the Hydro Outlet

Adult fish migrating upstream will be attracted to the flow of water where the tailrace returns water to the river. Fish may enter and swim against this flow and they could become trapped and consequently vulnerable to predation, or they may be killed in the turbine. The Fisheries Act requires that a fixed grating with minimum spacing between bars of 2.5cm should be installed across the point where the turbine exhaust water returns to the river. It is essential that, where possible, these gratings should be set parallel to the river flow, since this will encourage the fish to move upstream away from the turbine. Checks should be carried out by the hydro operator to ensure that the screens are working effectively:

Table 40: Small hydro projects golden rules.	
Top Tips when Considering a Hydroelectricity Installation	**Things to Avoid when Considering a Hydroelectricity Installation**
Contact the relevant water authority as soon as possible to obtain abstraction licence and permissions.	Do not overestimate the output from the hydro plant, as this will obviously impact revenue if it is inaccurate.
Make contact with the local angling groups to discuss the mitigation strategy for fish.	Do not underestimate capital costs for the project.
Use a reliable hydro consultant – engage them early in the project to plan and manage the installation and prepare the planning application/acquire abstraction licence.	Maintenance for these schemes is critical – debris clearance and inlet channel upkeep is essential.
Meet with the planning authority and the river management agency early in the project.	Do not forget to contact the DNO (electricity company) to get an accurate connection cost for the scheme. Connection cost could be the largest outlay in the project.
Get the river hydrology data going back over as many years as possible. An accurate forecast of river flow is essential for revenue estimates.	Do not forget that the river system is constantly changing and that drainage patterns and river flows may have changed significantly over the last couple of decades.

- Screens should be checked regularly for damage or obstruction.
- Screens should be kept clean.
- Screens should cover the full width of the tailrace.
- Replacement screens should be available in case of damage.
- Bypasses should be kept clear of debris to allow the unhindered passage of fish.

Electric Fish Barriers

Electric fish barriers have proved effective in preventing fish from entering tailraces. A typical fish barrier consists of a series of vertical electrodes 15 to 30cm apart, strung across the outlet. The electrodes have alternating electrical polarity, and when energized they create a localized electric field that is repellent to fish. Large fish tend to sense the field some distance from the barrier and avoid it, while smaller fish may make a closer approach. Electric barriers are not effective at turbine intakes and therefore are not recommended. When using electric fish barriers it is essential to take public safety into consideration at all times.

HYDROELECTRICITY GLOSSARY

Dam An obstruction, weir, dyke, sluice, embankment or structure built or placed in, or in connection with, any river for the sustaining of water for any purpose.

DNO (Distribution Network Operator) The energy or electricity company that operates the electricity grid. The DNO provides connection agreements and connection cost quotations, and can organize to have the work completed.

Draft tube The tube that contains the turbine runner and through which the water flows.

Energy output (kWh) The electrical energy output over a period of time (e.g. a year).

Fish farm Any undertaking for the culture of fish operating under the authority of a Fish Culture Licence.

Fish pass A channel for the free run or migration of fish in, over, or in connection with an obstruction in a river, lake or watercourse, including a fish ladder or any other contrivance that facilitates the safe passage of fish.

Inundation The flooding or submersion of land.

Lade, layde A man-made channel that conveys water from, and returns water to, a river for the purpose of supplying a mill or hydroelectricity installation. This includes the terms head race, which conveys water from the dam to the hydro plant and tail race, which conveys water from the hydro plant to the river.

Life-cycle analysis (LCA) A detailed analysis of the complete manufacture, installation, operation and decommissioning of the plant, with consideration for the environmental and energy balance over the lifetime of the system.

PLC A programmable logic controller is a microprocessor device that controls the start-up, on-load power output and shutdown sequences of the turbine.

Power factor correction A means of controlling voltage local to the turbine point of connection, using capacitors.

Power output (kW) The instantaneous electrical output from the hydro turbine.

Pump storage The use of stored water to manage electricity demand. At times when surplus electricity is available, it is used to drive pumps that raise water to a reservoir. This water is released through hydroelectric generators when the electricity has its greatest value.

Regulation The control of the turbine or guide vanes to optimize performance and output power.

Rotor The rotating element or shaft of the turbine.

Salmonid Fish of the salmon, trout or eel species.

Sluice Large gate valves or doors that can be opened or closed to admit water to a hydro plant.

Stress failure A failure mechanism in metals caused by vibration or cyclic stressing.

Trout Includes all fish of the brown trout kind, with the exception of sea trout and rainbow trout.

Watercourses Streams, rivers, canals and waterways.

USEFUL HYDRO CONTACTS AND WEBSITES

British Hydropower Association
www.british-hydro.org
12 Riverside Park, Station Road, Wimborne, Dorset BH21 1QU
Tel: 01202 880333 Fax: 01202 886609

Centre for Ecology & Hydrology
www.ceh.ac.uk
CEH Wallingford, Maclean Building, Benson Lane, Crowmarsh Gifford, Wallingford, Oxfordshire OX10 8BB
Tel: 01491 838800 **Fax:** 01491 692424
Email: wallingford@ceh.ac.uk

Department of Culture, Arts and Leisure (Northern Ireland)
www.dcalni.gov.uk/index/inland_fisheries.htm
Inland Fisheries Branch, 3rd Floor, Interpoint, 24–26 York Street, Belfast BT15 1AQ, Northern Ireland

The Office of Public Works (OPW)
www.opw.ie
Holds hydrometric data for rivers in the Republic of Ireland.

USEFUL PUBLICATIONS

Beach, M.H., *Fish Pass Design: Criteria for the Design and Approval of Fish Passes and Other Structures to Facilitate the Passage of Migratory Fish in Rivers*, Fisheries Research Technical Report no. 78 (Ministry of Agriculture, Fisheries and Food, 1984)

Hydropower: A Handbook for Agency Staff (Environment Agency, 2003)

O'Neill, G., and S. Connor, Protecting Fish: Guidelines for Water Abstractors, Inland Fisheries Advisory Handbook no. 1 (Department of Culture, Arts and Leisure [Northern Ireland], 2003)

Small Hydro-Electric Schemes: Impact on River Fisheries in Northern Ireland (Department of Enterprise, Trade and Development, 2001); available at www.detini.gov.uk/cgi-bin/downutildoc?id=52

Turnpenny, W.H., G. Struthers and K.P. Hanson, *A UK Guide to Intake Fish-Screening Regulations, Policy and Best Practice: With Particular Reference to Hydroelectric Power Schemes* (Department of Trade and Industry, 1998); available at www.berr.gov.uk/files/file15347.pdf

Republic of Ireland Legislation

Fisheries Amendment Act (1997)
Sea Fisheries & Maritime Jurisdiction Act (2006)

USEFUL WEBSITES

Department of Public Enterprise
www.irlgov.ie/tec/energy
Information about electricity supply in the Republic of Ireland.

European Small Hydropower Association (ESHA)
www.esha.be

International Hydropower Association
www.hydropower.org

Irish Hydropower Association
www.irish-hydro.com

Renewable Energy Information Office (REIO)
www.irish-energy.ie/reio.htm
Website supplies lists of hydro consultants and suppliers in the Republic of Ireland.

CHAPTER 8

Renewable Energy Technologies for the Near Future

While renewable energy technologies have been in everyday usage for a long time (*see* Chapter 1), there are a number of resources that have yet to be developed to their fullest potential, and these could be very meaningful indeed.

The heat energy emanating from the Earth's core, or deep geothermal energy, has been exploited widely, but so far there have been few attempts to tap into its potential in the British Isles. Some useful data has been gathered showing that there is scope for significant uptake, although more work is required to fully quantify the opportunities.

We live on islands, and the power of the seas that surround them – our marine energy – is clearly immense. Crashing waves bombard our vast coastline constantly and predictable tidal flows move huge quantities of water twice a day, but these natural resources are presently latent as a useable energy resource. Interest in wave power surged during the energy crisis of the 1970s, spawning a rush of technological developments, although few conversion devices have emerged as economically viable. Tidal energy has more recently come to centre stage and several developers are proceeding with designs that might eventually provide a significant contribution to the electricity demand of these islands. The timescale for the full commercialization of marine technologies is somewhere within ten to twenty years. Carbon reductions and impending renewable energy targets will be achieved using the established and commercially viable technologies – onshore and offshore wind – with a smaller contribution from a range of other technologies. The following brief review looks at three technologies that have an established theoretical resource and which could provide significant amounts of renewable power: geothermal energy, wave power and tidal energy.

DEEP GEOTHERMAL ENERGY

Parts of Great Britain and Ireland have a good supply of accessible underground hot rocks and hot water, and this makes up our deep geothermal energy resource. The resource occurs as heat in the underlying rocks up to 5km below the surface and as hot water in underground flows known as aquifers. In volcanic areas such as Iceland and New Zealand, molten rock can be very close to the surface, although this is not the case in the British Isles. Geothermal energy is available as a result of heat percolating up from the Earth's mantle into the rocks and ground waters above. Ground water is regarded as a shallow geothermal resource if it lies close to the surface and has a temperature within the range 10°C to 13.5°C. The relationship between the temperature and the underlying geology is complex and depends on a combination of factors, including rock type, the geological setting, the type of sedimentary basin and the occurrence of aquifers. In general, the temperature rises one degree Celsius for every 36m below the Earth's surface.

A History of Geothermal Energy Use

Geothermal energy manifests itself as steam geysers and hot springs, and these have been used for centuries across the world for bathing and heating. The Romans had hot spring baths in Britain but it was not until the twentieth century that geothermal power started being used to generate electricity.

The first geothermal electricity-producing power plant was demonstrated in July 1904 by Prince Piero Ginori Conti, who tapped into the resource at the Larderello dry steam field in Italy to power five light bulbs. In 1922 the first geothermal heating application in the United States was introduced by John D. Grant at the Geysers Resort Hotel, California. Grant eventually found enough steam in his boreholes to power an electricity generating station that provided lighting for the whole town. The first modern electricity-generating geothermal power plant appeared in 1960, when Pacific Gas and Electric (PG&E) began operation of the Geysers Power Station. It continued to generate for more than thirty years and had a peak output of 11MW. The largest group of geothermal power plants in the world is now located in California. In Europe there are a number of geothermal power stations: Iceland, for example, generates significant quantities of electricity from this resource.

Geothermal Resource

The low-grade geothermal heat resource accessed in the soils and rocks a few metres below the surface is, in fact, stored solar energy (*see* Chapter 3). This section looks at the technologies associated with the higher temperature resource. In deep geothermal energy, heat is extracted from deep within the ground and is used to provide space and water heating; if temperatures are high enough, it can also be used to generate electricity. The temperature at the Earth's centre is around 6,000°C, hot enough to melt rock, but the deep geothermal resource is drawn from the heat of the magma, the layer of molten rock lying just a few kilometres below the Earth's surface in some places. This heat produces hot ground water flows and it is also available in the hot rocks themselves. At 5km down, the temperature can be in excess of 200°C.

Aquifers

Aquifers are naturally occurring underground reservoirs of hot water. It is possible to pump the hot water to the surface through a drilled borehole and extract its heat. In an open system the water is discarded once the heat has been removed. Where the borehole is close to the sea, the cooled water can be dispersed into the sea. Normally, to preserve the resource, the water is returned to the aquifer via another borehole. The water from a borehole is frequently rich in minerals, making it brackish (salty). Aquifers can supply hot water for space heating in offices or large horticultural glasshouses, as well as for industrial processes, and most have a lifespan of around twenty to thirty years.

In the 1970s the United Kingdom developed a research and development programme to identify deep aquifers that produce water hot enough to be used directly in heating schemes. The programme identified fifty sites that coincided with urban centres (it is uneconomic to transport the water more than 30km). The UK government's Energy Technology Support Unit (ETSU) concluded that these sites could produce 650GWh per year by 2025. One site identified was Southampton and a demonstration scheme has been in operation since 1981 to prove the concept, pumping hot water more than 1.5km from an aquifer to the surface. The scheme has provided district heating for several buildings in the city, including the Town Hall, and produced up to 16GWh per year – enough for around a thousand homes.

Figure 88: The world's first geothermal electricity-producing power plant at Larderello in Tuscany, 1904. (Photograph courtesy the CSA Group)

Geothermal Electricity

Geothermal electricity can be generated using different types of power stations classified as dry steam, flash steam or binary-cycle, depending on the quality of the subterranean water and steam available. In dry steam power plants, steam (not water) is taken directly from the borehole, passed through a filter to catch rocks and debris, and then supplied to a steam turbine. The steam may also be passed through a heat exchanger to heat water for space heating. The steam must be purified to remove its hardness before it is used to drive a turbine, or the turbine blades will get furred up (as happens with a kettle or pipes in hard water areas), with disastrous consequences. Once the energy in the steam has been extracted, the condensed water is injected back into the ground. In some areas the natural supply of water producing steam from the hot underground deposits can become exhausted and it is then essential that replacement water is injected to replenish the supply. However, this technique requires careful management, since excess re-injection can cool the resource.

The most common form of geothermal power station in operation across the world today is the flash steam plant. It draws hot water from geothermal resources at temperatures above 180°C. The boiling point of water at the surface (at 1 Bar) is 100°C, but the pressure of the water several kilometres below the surface is high enough to prevent it turning to steam. When the water reaches the power plant at the surface, however, the drop in pressure causes it to flash into steam. This steam is used to drive a turbine and any water not flashed to steam is injected back into the underground reservoir.

The third type of geothermal power station is the binary-cycle power plant, which draws heat from a lower temperature resource, usually water at temperatures in the range 100°C to 180°C. In this case hot water comes to the surface, where it is passed through a heat exchanger containing heat pipes. The fluid inside the heat pipes boils at a lower temperature than that of the hot water. The heat pipes contain fluids such as iso-butane or iso-pentane and these are vaporized by the hot water. This vapour is used to power the turbine before it is condensed and returned to the heat pipe. Binary-cycle power plants have the advantages that they cost less and are more efficient, and

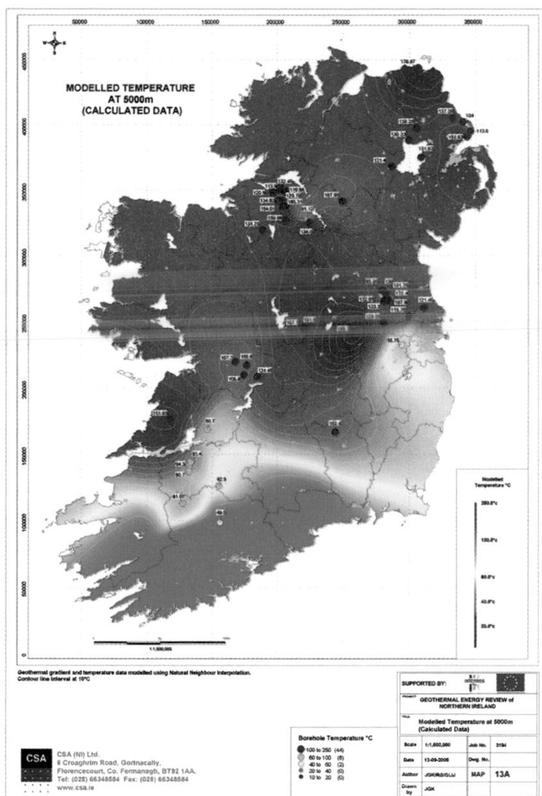

Figure 89: Geothermal map of Ireland at a depth of 5km, showing the temperature contours. (Image courtesy the CSA Group)

consequently they are becoming the most popular type of geothermal plant. Binary-cycle plants have no harmful emissions and, because they use heat pipe fluids with a lower boiling point than water, they are able to utilize lower temperature reservoirs, which represent a much larger resource.

Rocks that are hot but do not have an associated water or steam resource offer another type of power plant, the hot-dry-rock (HDR) geothermal power plant, which requires water to be pumped down into the underlying rock strata. The key characteristic of a HDR system (also called an Enhanced Geothermal System or EGS) is that it extends down into hard rock to at least 10km. At a typical site, two holes are bored some distance apart and the rock between them is fractured. Water is then pumped down the injection borehole (or injection well) and steam comes up the other borehole. This system has many advantages since it can be used anywhere, not just in tectonically active regions. However, HDR geothermal power stations require deeper drilling than the other types of geothermal power plant.

Advantages of Geothermal Energy over Fossil Fuels

Geothermal energy is classed as a renewable resource and it offers advantages over fossil fuels. The energy recovered from the rocks is clean and safe for the surrounding environment, and the technology is considered to be sustainable because the hot water used in the geothermal process can be re-injected into the ground to produce more steam. Geothermal power plants are unaffected by changing weather conditions and they can work continually throughout the year, providing baseload power. The economic viability of these power plants is good in so far as the fuel cost is zero once the plant has been constructed and they reduce reliance on fossil fuels. They also indemnify against fossil fuel price volatility. Around the world, geothermal energy plants have been constructed to serve a range of heat loads: a large geothermal plant can power entire cities while smaller power plants can supply more remote sites, such as rural villages.

Environmental Aspects of Geothermal Energy

From an environmental perspective, there are a few issues worth mentioning. Land stability in the area surrounding the plant can be affected during construction, particularly with an HDR system, which introduces water into hot dry rocks where there was none previously. Apart from the emissions due to fabrication and construction, some dry steam and flash steam power plants also emit low levels of entrained carbon dioxide, nitric oxide and sulphur (within the water and steam), although at much lower levels than those emitted by fossil fuel-burning power plants.

Geothermal plants usually include emission control systems that can inject these gases back into the ground, thereby reducing the carbon dioxide emissions to the atmosphere to less than 0.1 per cent of those from fossil fuel power plants. Geothermal sites are capable of providing heat for many decades, although eventually some locations cool down or the supply of water or steam can become exhausted. This occurs when the system, as designed, is too large for the site, since there is only so much energy that can be stored and replenished in a given volume of rock. Geothermal locations can undergo depletion and this leads to the question of whether or not geothermal energy can be regarded as truly renewable. However, over the course of time these locations will eventually recover some of their lost heat from the mantle's heat reserves.

Geothermal Potential

There are currently around 100GW of non-electricity producing geothermal heat plants (including ground souce heat pumps) installed across seventy countries worldwide. Estimates of the accessible worldwide geothermal energy resource vary considerably: they could amount to between 65GW and 138GW of electrical generation capacity using HDR technology. For HDR, the technological challenges are to drill wider boreholes and to break underground rock over larger volumes.

The geothermal resources in Britain are represented by low-grade heat reserves in deep sedimentary basins and by hot dry rock heat-stores that reside in radiothermal granites, with some in deep basement rocks where they are overlain by thick sediments. The low-grade heat resources are in Permo-Triassic sandstones at temperatures in excess of 40°C. To date, four deep exploration wells have been drilled to investigate the potential of these sedimentary aquifers. The principal HDR resource potential is associated with major granite deposits in south-west and northern England, where temperatures are predicted to be 200°C, at about 5.4km and 6km respectively. The HDR potential has been investigated by the Camborne School of Mines at their test site in Cornwall, where three boreholes were drilled to depths of between 2km and 2.5km.

According to the British Geological Survey, the HDR accessible resource at temperatures of more than 100°C and depths of less than 7km is 36 × 1021 Joules (equivalent to 130 × 104 million tonnes of coal). The low-grade geothermal resource above 40°C in the Permo-Triassic sandstones is 200×10^{18} Joules (equivalent to about 8,000 million tonnes of coal). These resources, if they could be developed, clearly represent a very significant potential in terms of the UK's energy requirement.

In 1986 the then Department of Energy ceased supporting the development of aquifers and concentrated on monitoring the Southampton scheme. It produced a refined estimate of the UK resource and its economics, noting that 'geothermal aquifers constituted a very limited energy resource and that commercial exploitation was unlikely'. In 1990 the Renewable Energy Advisory Group concluded that, 'the fundamental technical difficulties made the commercial development of HDR unrealistic in the foreseeable future'. Since then the British government has collaborated with the European HDR programme and work is taking place at UK, French and German sites.

Recent studies have shown that in Ireland there is a significant resource at depths around 3km to 5km to provide sufficient geothermal energy for a small power station. A similar resource in Germany has been used to build the 100MW Unterhaching power station, near Munich, which has been operating since 1998.

WAVE ENERGY

The energy in the tides and waves around Britain and Ireland has long been recognized as a fantastic resource. As we move into a sustainable energy future, developing marine technologies offer new opportunities to harness the energy of the tides, waves and offshore winds. Wave energy exists as a result of the wind, which is ultimately a form of solar energy; it has significant diurnal and seasonal variability and can change dramatically from one location to another around the coast. The idea of extracting energy from the waves is not new and useful advances were made by Japanese scientists in the mid-1940s. The real impetus to develop this resource arrived with the oil crisis of the 1970s, when countries that depended heavily on oil realized that, if they were to secure their energy future, they needed to look away from a reliance on fossil fuels. Throughout the 1970s and 1980s research in various countries including Norway and the UK brought forward a number of wave energy devices to generate electricity. Despite the obvious potential, these early machines struggled to reach commercial maturity, although wave energy has been used to power navigation and weather monitoring buoys. Several larger devices are now approaching the commercial demonstration phase.

The wind's movement over the seas exerts a frictional stress on the water's surface, resulting in the formation and growth of waves. Once these nascent waves reach a certain size, the wind can exert a strong lift force on the upwind face of the wave to further increase its height.

Globally, the best sites for capturing wave power are the north and south temperate zones: the prevailing westerly winds in these zones blow strongest in winter. Britain and Ireland lie well within these areas. The exposed west-facing coastlines of these islands offer the best opportunities, particularly the west coast of Ireland and Scotland. Estimates put the European resource at around 400TWh, which represents 17 per cent of the EU demand for electricity. In the UK the theoretical resource is substantial and is projected to be around 61 to 87TWh, which could satisfy 20 per cent of UK electricity demand, if it could be harnessed practically.

Wave Energy Conversion (WEC) Technology

Despite the fact that the power in the waves is patently obvious, wave energy is difficult to harness and convert into meaningful amounts of electricity. WECs require a system of reacting forces in order to extract the energy from the waves and this is one of the biggest design challenges. To create such a system, two or more parts of the converter need to move relative to each other, while at least one part interacts with the waves. Some devices allow one component to move freely with the waves, while another is held fixed (an example is a floating buoy that reacts against the seabed).

Alternatively, all of the components may move relative to one another. The various components may be described as reactors or displacers, depending on their function in the conversion process. A displacer may be a column of water or the floating chamber of a buoy. The reactor may be a component attached to the seabed or the seabed itself. It might also be another structure or component that is not fixed but which moves to create a reaction force, for example, by moving at different times or by a different amount.

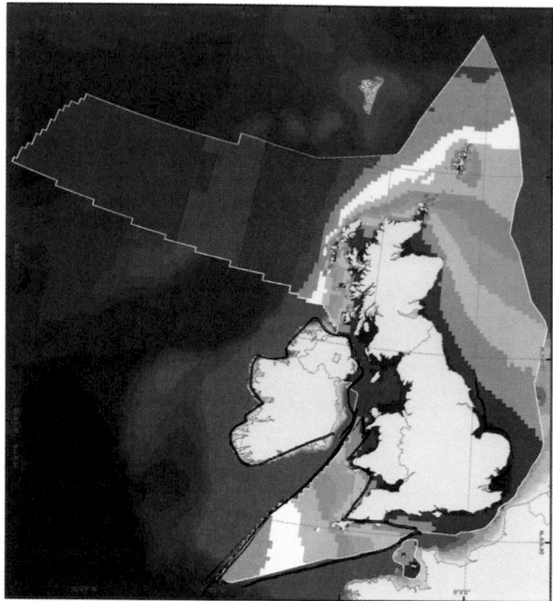

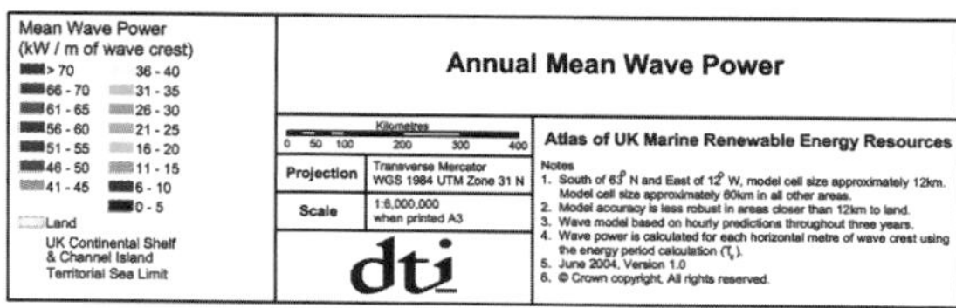

Figure 90: Annual mean wave power in the seas around the United Kingdom. (Image courtesy of the Department of Business, Enterprise and Regulatory Reform)

The essential requirement for any WEC is that it must survive in the marine environment. This means that it should be designed to cope with fatigue stress and avoid extreme destructive loading. To be most effective, wave energy conversion needs to generate electricity from wave motion. At the point of capture, wave power is available from WECs at low speeds and it involves large forces and movement that is not in a single direction. Conventional electricity generators are designed to operate at higher speeds (1,500–3,000rpm) with lower input forces and they prefer to rotate in one direction only.

Wave energy converters can harness wave energy in three areas: on the shoreline, close to shore (or nearshore) and offshore. Nearshore devices may be bottom-fixed or floating. Embedded devices are built into the shoreline and they use the force of the waves to push or rotate a mechanical coupling to drive a generator. There are a wide variety of designs using the interaction between the displacer and reactor components of a converter.

The power take off (PTO) from the WEC can be through the use of hydraulic, mechanical, pneumatic or electrical transducers. Hydraulic rams, elastomeric hose pumps and pump to shore systems, as

Marine Energy Terms

Sea state In real sea conditions, many wave heights and periods occur simultaneously and it is necessary to resort to a statistical description. Data for a particular site can be summarized on a scatter diagram, which is a record of the wave motions showing the number of occurrences of particular combinations of significant wave height (Hs) and zero up-crossing period (Tz). Each combination of Hs and Tz is referred to as a sea state.

Wave energy The energy content of waves is a function of both wave height and period.

Wave height A measure of the amplitude of oscillation of water particles in the vertical direction with respect to a fixed point. Since the oscillations exhibit a range of values in real sea conditions, it is helpful to talk about a representative wave height. The significant wave height (Hs) is commonly referred to and is approximately equal to the average height of the highest one-third of the waves. (The actual definition is four times the root mean square of the water levels relative to the mean water level. The mean water level is the level the sea would be at if it was calm).

Wave period The time that elapses between successive peaks or troughs of a wave passing a fixed point. Like wave heights, waves exhibit a range of periods and a representative term is helpful. The zero up-crossing period (Tz) is one such term and is the average time interval between successive crossings of the mean water level in an upward direction.

Sources: The Carbon Trust (www.thecarbontrust.co.uk/technology/technologyaccelerator/ME_guide.htm)

British Wind Energy Association, Marine Renewable Energy (www.bwea.com/marine/index.html)

well as more conventional hydroelectric turbines and air turbines, are examples of PTO systems. Most converters utilize the waves' oscillating water mass as the source of power. Waves can be used to cause the water level in a chamber to rise and fall in a bobbing motion. This motion forces air in and out of a receiver, which can drive a turbine to provide electricity.

The short-term energy available in waves may be extreme, especially in storms, and conversion devices must be built to minimize storm damage and tolerate saltwater corrosion. Extreme loads may be more than a hundred times greater than the working average. Failure mechanisms include bearing seizure, broken welds and snapped mooring lines.

Economics and Environmental Impact of Wave Power

The economics of wave energy are improving, although machines designed to be located and operated remotely in the hostile ocean environment will, by their very nature, be costly. Developments over the last decade have provided significant improvements in the economic feasibility of wave energy, but the fabrication, installation, repair and maintenance costs will all need to be addressed if wave energy converters are to compare favourably with other sources of electricity generation. In the past, over-engineering to anticipate these challenges has significantly increased costs. For economic energy production, WECs should be designed to maximize the wave energy capture, to minimize the structural content and to reduce component redundancy. Losses incurred in the PTO system must be minimized and the device should be capable of ready mass production. Modular construction with condition monitoring would be essential requirements. In the coming decades effort will need to be channelled into developing new materials, introducing robotics and automation to reduce costs, and improve remote sensing to maximize production and reduce maintenance requirements.

From an environmental point of view there is minimal impact on marine flora and fauna and, apart from the construction and installation processes, there are no emissions of carbon dioxide during operation. Financial support mechanisms are not yet finalized, but it is proposed that marine technologies will receive enhanced support through the UK's Renewable Obligation. Ireland also plans to put in place strong financial support for their marine energy programme.

Wave Power Development

A number of governments across the world have invested money in wave power development and several commercial demonstration projects, using a range of technologies, are now operational or under construction.

Although the UK Wave Energy Programme was officially shut down in March 1982, funding has recently been announced for the development of a Wave Hub off the north coast of Cornwall. The hub will allow arrays of wave-energy generating devices to be connected to the electricity grid. Initially 20MW of capacity will be connected, with potential expansion to 40MW. It is anticipated that wave energy developers will be keen to trial their demonstration machines by connecting to the hub.

The energy potential in waves is greater than that in the tides, and wave power can be exploited at many more locations. The following examples indicate in more detail the current status of some demonstration wave power technologies. A full listing of demonstration and test wave energy devices is held by the European Marine Energy Test Centre (EMEC; *see* page 196).

Limpet

The world's first grid-connected, commercial-scale wave energy plant is Limpet, which is on the island of Islay, off the west coast of Scotland. Commissioned in November 2000, it is a shoreline wave energy converter utilizing an inclined oscillating water column with Wells turbine power take-off. The water depth at the entrance to the OWC is typically seven metres. The design of the air chamber is important to maximize the capture of wave energy and conversion to pneumatic power. The turbines are therefore carefully matched to the air chamber to optimize their power output. The performance has been optimized for annual average wave intensities of between 15 and 25kW/m. The water column feeds a pair of counter-rotating turbines, each of which drives a 250kW generator, giving an output of 500kW. The Limpet design makes it easy to build and install. Its low profile

Wave Energy Converters

Salter Duck In response to the oil crisis of the 1970s, Professor Stephen Salter of the University of Edinburgh led the development of wave energy with his 'Edinburgh Duck', which still represents a benchmark for wave energy performance. The duck bobs up and down on waves and a special turbine converts this movement into electricity. Prototype Salter Ducks are as large as double-decker buses and they need to be arranged in set patterns in the sea, to take advantage of wave formations.

Oscillating Water Column (OWC) Devices OWCs use the periodic rise and fall of the waves to drive air through a turbine and so generate electricity. OWC devices can be offshore, nearshore or shoreline, and they have proven enormously successful in wave energy conversion designs. However, they have some drawbacks. For shoreline devices that use enclosed partially submerged structures, the capital cost of the barrier wall and enclosure is high and accessibility for heavy equipment is difficult. Shoreline OWCs require an oceanfront site and these may be restricted for environmental and planning reasons. Coastlines are frequently classified as areas of outstanding natural beauty (AONBs), or as areas of special scientific interest (ASSIs) for marine flora and fauna protection. Nearshore and offshore OWCs do not suffer these disadvantages but they are obviously more difficult to access and maintain.

Wells Turbine Wave energy converters operate in a highly corrosive saltwater environment and this can lead to serious problems when the water is in contact with mechanical parts over long periods. To overcome this, the oscillating water column design (described above) was developed to use pressurized air instead of seawater. The OWC produces an oscillating bi-directional flow and this must be managed. The Wells turbine was designed by Alan Wells from Queen's University, Belfast, in the late 1980s to overcome these difficulties. The Wells turbine can accept bi-directional flow and produce turbine rotation in one direction only, regardless of the direction of the flow of water or air. This was achieved through the design of the blades, which are similar to an airfoil except that they are symmetrical about the horizontal axis. Normally airfoils are elliptical in shape and asymmetrical. A conventional airfoil utilizes the lift force provided, but the Wells turbine uses the lift and drag force to ensure that the turbine blades continue to rotate in the same direction. One drawback of the Wells turbine is the aerodynamic loss in efficiency. Small, weak waves and large storm-force waves produce inefficient running, so the Wells turbine may only be suitable within a restricted set of wave heights.

TAPCHAN Tapered channel (TAPCHAN) wave energy converters were originated and constructed by Norwegian researchers in 1985. In this design, the front of the wave converter faces the ocean and it is surrounded by high hemispherical concrete walls on either side. The wave entrance to the structure is at a slight incline on the seaward approach, with a reservoir on the shore side. The channel is very wide closest to the sea, and tapers to a smaller throat as it approaches the reservoir. The reservoir fills as the water rushes shoreward, passing through a turbine inlet pipe. The kinetic and potential energy of the wave rushing through the inlet pipe spins the turbine and the generator, to provide electricity. TAPCHAN requires excellent average wave energy at a site in order to exert enough force to push the water into the reservoir.

Overtopping WECs These are structures over which the waves topple into a reservoir. The collected seawater provides a head to drive hydro turbines, which produce electricity as the water flows back out to the sea from the bottom of the reservoir.

Point absorber WECs Point absorbers are floating structures that absorb energy from all directions through their movements on or near the water surface. They can be designed to amplify the wave movement to maximize the amount of power available for capture. Examples of point absorber WECs are Oyster or Energetech OWC.

Terminator WECs These are structures that move at or near the water surface, absorbing energy in one direction only. The converter is placed at right angles to the wave direction so that, as waves arrive, the device restrains them. Again, amplification can be used and the PTO can use a range of techniques to extract the energy. An example of a terminator WEC is Limpet.

Wave Energy Converters *continued*

Attenuator WECs These are long floating structures similar to the terminator, but orientated parallel to the wave direction rather than at right angles to it. They ride or cut through the waves like a ship and movements of one section of the device relative to another can be restrained so as to extract energy. A theoretical advantage of the attenuator over the terminator is that its area at right angle to the waves is small, and therefore the forces it experiences are much lower. An example of an attenuator WEC is Pelamis.

Wave Rotor This is a turbine design turned directly by the waves working on the principle of differential pressure. It is coupled to a generator to produce electricity.

Other devices use contained volumes of water, or exploit differences in water pressure, such as the Wave Dragon.

Sources: The Carbon Trust (www.thecarbontrust.co.uk/technology/technologyaccelerator/ME_guide.htm)

British Wind Energy Association, Marine Renewable Energy (www.bwea.com/marine/index.html)

against the shoreline gives Limpet low visibility, so it does not intrude on coastal landscapes or views.

PowerBuoy

The Pacific Northwest Generating Cooperative in the United States is funding the building of a commercial wave power park at Reedsport, Oregon. The project will utilize the PowerBuoy technology, which consists of modular, ocean-going buoys. The rising and falling of the waves moves the buoy-like structure, creating mechanical energy that is converted into electricity and transmitted to shore over a submerged transmission line. A 40kW buoy has a diameter of 4m and is 16m long, with approximately 4m of the unit rising above the ocean surface. Using the three-point mooring system, it is designed to be installed up to 8km offshore in water between 30m and 60m deep.

Oyster

The Oyster is a seabed-mounted point absorber converter in which the PTO is through the use of hydraulic transducers. The nearshore device, designed by Professor Trevor Whittaker from Queen's University, Belfast, operates in 10 to 15m depth of seawater.

Pelamis

One of the most successful implementations of the surface-following device is the Pelamis wave energy converter. Sections of Pelamis oscillate with the movement of the waves, each resisting motion between it and the adjoining section, forcing pressurized oil to drive a hydraulic ram, which in turn drives a hydraulic motor. The Aguçadoura wave park, near Póvoa de Varzim in Portual, will initially use three Pelamis P-750 machines to generate 2.25MW of electricity. Plans for a wave farm in Scotland with four Pelamis machines were announced in February 2007.

Wave Dragon

The Wave Dragon overtopping offshore energy converter employs large 'arms' to focus waves up a ramp into an offshore reservoir. The water returns to

Figure 91: The Limpet wave energy converter, operating on the island of Islay since 2000. It generates 500kW of electricity. (Image courtesy of Wavegen)

Figure 92: In PowerBuoy, the rising and falling of the waves offshore causes the buoy to move freely up and down. The resultant mechanical stroking is converted via a sophisticated power take-off to drive an electrical generator. (Image provided courtesy Ocean Power Technologies, Inc)

the ocean under gravity, passing through hydroelectric generators.

AquaBuOY

In the AquaBuOY offshore wave energy device, the vertical movement of the wave drives two-stroke hose pumps to produce pressurized seawater, which is then directed into a turbine connected to an electrical generator. The power is transmitted to shore by means of a secure undersea transmission line. A commercial wave power production facility utilizing the AquaBuOY technology is under construction in Portugal.

CETO

A device called CETO is under test in Fremantle, Western Australia. This uses a seafloor pressure transducer coupled to a high-pressure hydraulic pump. The wave energy converter is used to pump water to shore for driving hydraulic generators which produce electricity, or for running desalination plants.

Also in Australia, a device installed near Wollongong, New South Wales, uses a parabolic reflector to concentrate wave energy into an oscillating water column. Air is driven through a Denniss-Auld turbine, which, like the Wells turbine, is designed to rotate in a constant direction in the oscillating airflow.

Neo-AeroDynamic

This is an airfoil-based design that harnesses the kinetic energy of wave-induced flow through an artificial current directed around its centre. The device differs from others in its ability to transfer wave power directly into rotational torque to drive a generator. Neo-AeroDynamic has a high-energy conversion efficiency and has applications in other renewable technologies.

TIDAL ENERGY

The interaction between the gravity of the Earth, Moon and Sun gives rise to enormous tidal energy in the oceans and seas. This is an inexhaustible energy source and can be classified as renewable. In Europe, tide mills have been used for more than a thousand years, mainly for grinding grain, and their remains can be found at sites across the UK and Ireland. The movement of water caused by the ebb and flow of tidal currents (tidal stream), or from the rise and fall in sea levels between high and low tides (tidal range), is utilized to tap into the massive resource. The energy supply is clearly available, reliable and plentiful, but converting it into useful electrical energy is difficult and expensive. Tidal power can be considered as being of two main types:

Tidal Stream. Tidal stream systems use the movement, principally the kinetic energy, of the huge quantities of flowing water to power turbines. Early indications from demonstration sites suggest that tidal stream devices can have minimal environmental impact.

Tidal Range. Energy from tidal range can be captured using tidal barrages or tidal lagoons, which make use of the difference in height, the potential energy or head, between high and low tides. Huge civil infrastructure costs, the lack of viable sites and concerns around environmental issues currently restrict the potential deployment of tidal range schemes around the world.

Tidal Stream Energy Converters

The concept behind tidal stream generators is very straightforward: they utilize the energy from marine currents in much the same way as wind turbines draw their energy from the mass movement of air. To provide meaningful output in areas where sufficiently rapid tidal flows occur, tidal stream turbines need to be constructed in arrays, analogous to wind farms under water. Locations for potential sites are where natural flows are concentrated, such as the west coast of Canada, the Straits of Gibraltar, the Bosphorus, and at a number of locations in South-east Asia and Australia. These flows occur almost anywhere that has entrances to bays and rivers, or between land masses where water currents are constrained. There are a range of opportunities in the waters around the British Isles, especially in the Irish Sea. Many of the potential sites are close to population centres and this may be important for the economics. Water has a density more than 800 times that of air, which means that the tidal flow energy density is much higher. Fast tidal currents of around 3m/s to 4m/s are available and these represent the best locations.

In Europe there are several commercial prototypes that have shown promise. Trials in the Straits of Messina between Italy and Sicily started in 2001. The Australian company Tidal Energy Pty Ltd undertook successful trials of highly efficient shrouded turbines (see below) on the Gold Coast region of Queensland in 2002, and it is presently considering opportunities in Asia and Europe. There are similar developments occurring in the USA, where Verdant Power operates a prototype project in the East River, between Queens and Roosevelt Island, New York City. The world's first grid-connected prototype turbine, which is capable of generating 300kW, started operation in November 2003 in the Kvalsund, south of Hammerfest, Norway. The Norwegians propose to install a further nineteen similar turbines.

Excellent tidal runs in the British Isles include the Severn, Dee, Solway and Humber estuaries. Other sites around the British Isles include parts of the Irish Sea, areas to the north of Anglesey, to the east of the Copeland Islands, in the North Channel at Torr Head and to the south of Rathlin Island. During 2003 a 300kW Periodflow marine current propeller type turbine was tested off the coast of Devon, and a 150kW oscillating hydroplane device, the Stingray, was tested off the Scottish coast. SeaGen, a 1.2MW turbine manufactured by Marine Current Turbines Ltd (MCT), was installed in Strangford Lough Narrows, Northern Ireland, in April 2008 before connection to the grid. In November 2007 MCT signed an agreement with Canada's Maritime Tidal Energy Corporation to have tidal stream turbines operating in the Bay of Fundy, Nova Scotia, by 2009.

Shrouded Tidal Turbines

A shrouded tidal turbine has the turbine enclosed in a venturi-shaped shroud or duct. This creates an area of low pressure behind the turbine, allowing it to operate at higher efficiency and with a power output typically three to four times above a turbine of the same size. The shrouded design permits a smaller turbine to be used at sites where it would be difficult to install large turbines. They can be assembled in greater numbers as an array, stretching across a seaway or a river. They are straightforward to connect to the grid and may be used to support the power system in remote community projects. The shroud design allows sites with low

Tidal Cycles

The tides are caused by the gravitational influence of the Moon and the Sun acting upon the oceans of the Earth as it rotates. The Moon exerts approximately twice the tidal force of the Sun. The relative motions of these bodies cause the surface of the oceans to be raised and lowered periodically, according to a number of interacting cycles. These include:

- **A daily or half-daily cycle**, owing to the rotation of the Earth within the gravitational influence of the Moon. This produces the familiar high and low tides, which are experienced at different times of the day depending on location. In the British Isles, high and low water usually occurs twice daily (it is semidiurnal), with the time of high water advancing by approximately 50 minutes per day.
- **A worldwide 29.5-day cycle**, resulting from the relative position of the Moon and Sun. This results in spring tides and, seven days later, neap tides. Spring tides are those half-daily tides with the largest range (i.e. highest high water and lowest low water), while neap tides have the smallest range. Spring tides occur shortly after the full and new Moon, with neaps occurring shortly after the first and last quarters. For any given location, the spring tide high water will always occur at the same time of day.
- **A half-year cycle**, due to the inclination of the Moon's orbit to that of the Earth. This gives rise to the largest spring tides, around the time of the March and September equinoxes, and the smallest spring tides, approximately coincident with the summer and winter solstices.

There is also an **18.6-year tidal cycle** that results in larger than average tides, requiring estimations of tidal resource to be based on an average year. The range of a spring tide is commonly about twice that of a neap tide, whereas the half-yearly cycle imposes smaller perturbations. In the open ocean, the maximum amplitude of the tides is about one metre. Tidal amplitudes are increased substantially towards the coast, particularly in estuaries. This is mainly caused by shelving of the seabed and funnelling of the water by estuaries.

These factors combine with the Coriolis effect and frictional effects to make the tidal range and times of high and low water vary substantially between different locations on the coastline. This also results in a large variation in the energy that can be obtained from the tides on a daily, weekly and yearly basis.

Source: *Turning the Tide: Tidal Power in the UK* (Sustainable Development Commission, 2007)

tidal velocities, which were previously considered unfeasible, to be exploited. It can be integrated into a tidal fence or tidal barrage to enhance the performance of the turbines. Shrouds also form a protective shield for tidal turbines, blocking floating debris and eliminating bio-fouling, which can be a problem in shallow water. Higher water speeds through the turbine blast the shroud throat and turbine clean, so preventing bio-organism growth.

Shrouded turbines also have some disadvantages, since they need to be located below the mean low water level, which necessitates mono-pile fixing to the seabed. Shrouded tidal turbines are mono-directional and constantly need to face upstream, although they can be floated under a pontoon on a swing mooring, or fixed to the seabed. This means that they can be yawed like a wind turbine to continually face upstream.

Environmental Impact of Tidal Turbines

Tidal stream turbines have relatively slow rotational speeds, so they should have a minimal impact on and negligible interference with marine life and the environment, and they will have little, if any, visual impact. These aspects are under intense study at the various demonstration sites, particularly at the Strangford Lough Narrows, which is possibly one of the most environmentally sensitive areas in Europe. Tidal stream turbines present opportunities for remote communities, such as those on islands and rivers, especially where it is not feasible to connect them to grid supplies. They have the advantage of being cheaper to build and do not have the environmental problems associated with a tidal barrage. Tidal power stations require sites with good tidal stream velocities. These sites are limited in number and it may be necessary to legislate for them to ensure that they are developed using the most effective technologies.

A recent study of the long-term economic potential of tidal stream technologies in the UK was completed in 2006 by The Carbon Trust. The study concluded that, although costs will initially be higher than those of existing electricity generation, as the installed capacity increases the costs will fall, and eventually they will be comparable with the base price for fossil fuel-generated electricity, particularly if this price rises. The Carbon Trust estimated that 1GW to 2.5GW of tidal stream capacity could be installed across Europe by 2020, with the majority likely to be in the waters around the British Isles.

Figure 93: Artist's impression of an array of marine current turbines showing turbine raised above water level and maintenance access via boat. (Illustration reproduced courtesy of Marine Current Turbines Ltd)

Tidal Barrages

A tidal barrage is a huge dam built across a river estuary. As such, it incurs massive civil engineering costs and raises significant environmental issues during construction and operation. Barrages have tunnels through which water can pass with the ebb and flow of the tides. In the same manner as hydroelectric plants, the water movement through these tunnels can be used to power turbines and generate electricity. With a barrage in place, shipping traffic can still move up and down the estuary, subject to passing through lock gates that operate in the same way as those used on canals. Tidal power stations generate for around ten hours each day when the tide is flowing in or out, and, since the tides are totally predictable, other power stations can be scheduled to generate at times when there is no electricity from the barrage to ensure continuity of supply.

Components

Barrages are used to capture and contain the movement of water through the use of earthwork embankments and civil engineering structures known as caissons (massive concrete blocks). Other components, apart from the caissons, include sluices (large gates or valves), turbines and ship locks. Embankments seal a basin or basins where it is not blocked by caissons. The sluice gates used in tidal power applications are flap gate, vertical rising gate, radial gate and rising sector gate sluices.

Potential

There is a limit to the number of suitable tidal barrage sites across the world and there are only a few in the British Isles. In 1966 the largest tidal power station in the world, and the only one in Europe, was built across the Rance estuary in northern France. It has an installed peak capacity of 240MW and an annual production of 600GWh. The first tidal power site in North America was the 18MW Annapolis Royal Generating Station in Nova Scotia, which opened in 1984 on an inlet of the Bay of Fundy. The Soviet Union constructed a small 0.5MW barrage scheme at Kislaya Guba on the Barents Sea. China has developed several small tidal power projects, including one large facility in Jiangxia, and a tidal lagoon is scheduled to be built near the mouth of the Yalu.

In the UK plans for a Severn Barrage between Brean Down in Somerset and Lavernock Point in Wales have been developing over a number of years, with enthusiasm for the project waxing and waning. A barrage across the Severn Estuary would mean that the tides at Weston-super-Mare would not recede and some water would be retained in the estuary for most of the time. Such a barrage might have more than 200 hydroelectric turbines and provide around 8,000MW of power. It would take seven years to build and could provide

Figure 94: The SeaGen, a 1.2MW marine current turbine, being assembled in the Harland and Wolff shipyard prior to installation in the Strangford Lough Narrows. (Image by the author/www.marineturbines.com)

approximately 7 per cent of the electricity needs for England and Wales. Both Scotland and Northern Ireland have locations that would also be suitable for tidal barrages.

Generation

The key to barrage generation is to trap the rising (ebbing) water behind the barrage and then to release it back to the sea, generating electricity, when the tide has receded to lower the water level on the sea side.

During ebb generation, the sluices on the barrage and the tunnel inlet gates are opened to the tide and left open until high tide is reached, which ensures that the catchment basin is filled. The sluice gates are then closed. Pumps may be started to further raise the basin water level. The turbine gates remain closed until the sea level falls to create a head across the barrage. When the differential in heights is sufficient, the tunnel inlet gates are opened and water flow through the turbines starts electricity generation, which continues until the water level is the same across both sides of the barrage. The cycle is then repeated with the sluices being opened, the turbines shut down and the basin filled again.

Two-Basin Tidal Barrages

It is possible to operate a tidal barrage between two basins or tidal lagoons. Turbines are located within channels constructed between the basins. One basin is filled at high tide and the other is emptied at low tide. This offers advantages over normal schemes in that the time of generation can be adjusted with high flexibility and it is possible to generate almost continuously. Generally, two-basin schemes are very expensive to construct due to the cost of the extra length of barrage. Two-basin generation becomes more economically attractive when the natural geography is favourable.

Environmental Aspects of Tidal Barrages

Apart from the production of huge amounts of renewable energy, a barrage across the Severn Estuary, for example, would have a number of additional benefits. These include protecting a large stretch of coastline against damage from high storm tides and providing a ready-made road bridge. However, the significant changes to the flows in the estuary could have huge effects on the estuarine ecosystem and may cause pollution problems. For example, the Severn Estuary carries out to sea sewage and other wastes from several places including Bristol and Gloucester. A barrage would introduce issues in relation to water and sewage entrapment and would constrain the feeding habits of wading birds, which currently rely on the exposed mudflats.

The volume of water that enters a basin is reduced once a barrage is constructed. The turbidity of the water, that is the amount of matter in suspension, decreases because smaller volumes of water are exchanged between the basin and the sea. Lower turbidity allows sunlight to penetrate to greater depth, improving conditions for the phytoplankton. The changes propagate up the food chain, causing a general transformation of the ecosystem. Another influence is the salinity of the basin. When less water is exchanged with the sea, the average salinity inside the basin falls, which also affects the ecosystem. These changes will directly affect the flora and fauna, particularly the birdlife attracted to the basin.

Riverine estuaries usually have high volumes of sediment moving through them, from the rivers to

Power in the Tides

The energy content of tidal streams is dependent on the current velocity.

Current velocity is the speed of water particles moving in the tidal stream in the mean flow direction. Stream velocities above 3m/s are considered to be good for tidal stream development. In fast tidal runs velocities of 4 to 5m/s and above are experienced for short periods.

The power in a tidal current or stream is proportional to the cube of the water velocity (m/s). As is the case with wind energy, the cube relationship means that small changes in predicted tidal current velocity will lead to large changes in the predicted power density. It also means that there is approximately eight times more tidal stream power during spring tides than neap tides.

The tidal stream energy resource can be categorized in terms of its theoretical availability, its technical availability and its practical availability. For a tidal stream site, the spring tidal peak velocity relates to the theoretically available resource or the unconstrained resource. The area of water deep enough for the chosen technology relates to the technically available resource, and the potential impact on shipping lanes and general navigation relates to the practical available resource. Clearly, the practical resource may be several orders of magnitude less than the theoretical resource.

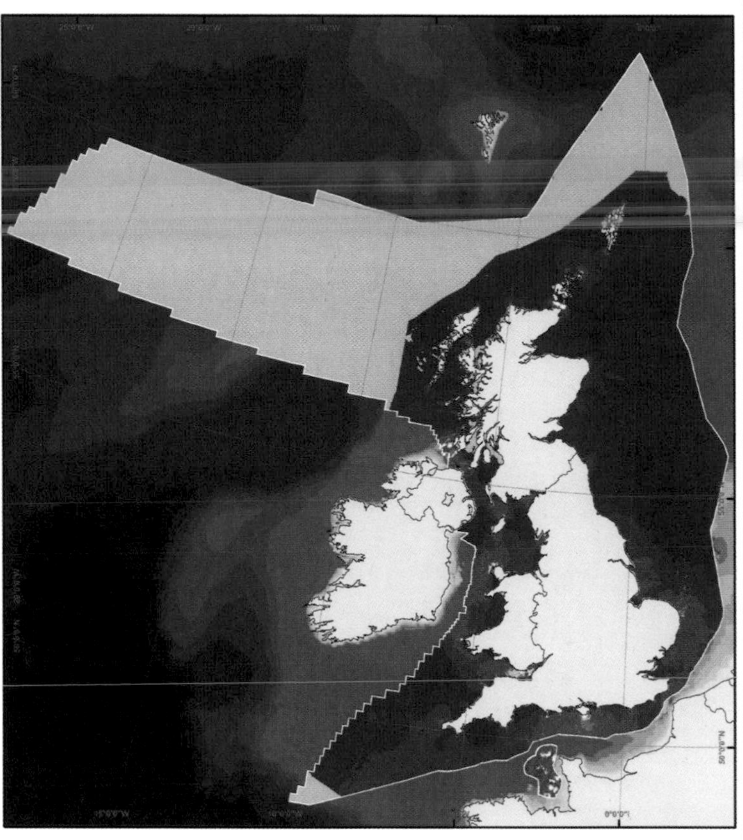

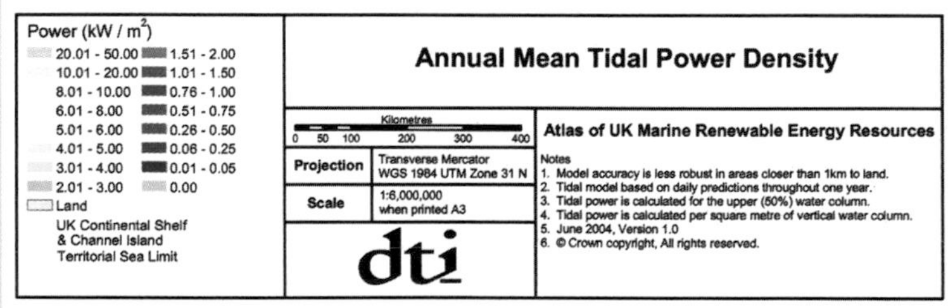

Figure 95: The annual mean tidal power density in the seas around the United Kingdom. (Image courtesy of the Department of Business, Enterprise and Regulatory Reform)

Source: *Turning the Tide: Tidal Power in the UK* (Sustainable Development Commission, 2007)

Table 41: Advantages and disadvantages of tidal energy.

Advantages	Disadvantages
During operation, tidal energy is free and it will be available for many years. Once constructed, it produces no greenhouse gases or other pollutants. Tidal energy consumes no fuel. Electricity is provided reliably. Maintenance is straightforward and inexpensive for barrages. Tidal flows are totally predictable. Within a decade, tidal stream turbines will be economically competitive with existing electricity generating technologies and they will have a negligible environmental impact.	A barrage across an estuary is very expensive to build. Tidal stream may disrupt marine flora and fauna. A barrage affects a very wide area – the environment is changed for many miles upstream and downstream. Wading birds rely on the tide uncovering the mud flats so that they can feed. There are few suitable sites for tidal barrages. Tidal energy can provide power for only around 10 hours each day.

the sea. The introduction of a barrage into this type of estuary may result in sediment accumulation within the basin, impacting the ecosystem and possibly constraining the operation of the barrage.

As with hydroelectric plants, careful consideration needs to be given to fish movement and protection during barrage construction and operation. Fish can move through sluices safely, but when these are closed fish will seek out turbines and attempt to swim through them. Also, some fish will be unable to escape the water speed near a turbine and will be sucked through. Even with the most fish-friendly turbine design, fish mortality may be a problem (from pressure drop and contact with blades). Designers of tidal barrages need to refine fish passage technologies (fish ladders, fish lifts, electric and sonic guidance) to minimize the impact of the barrage on fish life.

Any environmental evaluation needs to consider the influence of a range of variables, including alterations that affect:

- Water levels (during operation, construction, extreme conditions, etc.)
- Currents
- Waves
- Power output
- Turbidity
- Salinity
- Sediment movements.

Economics of Tidal Barrage Schemes

Although tidal barrage schemes have high initial capital costs, they benefit from low running costs. They also represent an efficient way of converting the potential stored and kinetic energy of the water into large amounts of electricity. Private investors have been reluctant to invest in these projects as a high capital investment is required up front, with low returns over long periods (perhaps 100 to 120 years).

Governments can accept much lower rates of return against major investments, but they have, so far, been unwilling to commit for the same reasons. In the UK, for example, the role of tidal energy has been recognized in a number of energy policy reviews and the government has expressed the need to consider and embrace the broader global and national goals of renewable energy, in approving tidal projects. The technical viability and siting options available are well understood, but so far the financial investment needed to bring forward a tidal barrage has not been realized. In November 2007 the Sustainable Development Commission (SDC) published a report on tidal energy, *Turning the Tide: Tidal Power in the UK*, which 'recognized the importance of giving serious consideration to a Severn Barrage, within a framework that places a high value on the long term public interest and on maintaining the overall integrity of internationally recognised habitats and species'.

Looking to the future, we are moving to a situation where renewable technologies will be more cost-effective, where fossil fuels will be less available, where there is likely to be limited nuclear energy deployment and where land-based renewables may be fully exploited. In this scenario, offshore wind, wave energy and tidal energy will be the obvious choices, although climate change itself will directly affect the development of these technologies, since the wave and wind climate and tidal current variation due to changing weather patterns will be more evident. There will be more intense storms (*see* Chapter 1), leading to higher operating loads for these technologies, and at coastal locations the results of any sea level rises will be important – especially for fixed wave energy converters and tidal barrages.

FUTURE TECHNOLOGY GLOSSARY

Coriolis effect Unrestrained objects on the surface of the Earth experience a Coriolis force, and appear to veer to the right in the northern hemisphere, and to the left in the southern.

Estuarine ecosystem The biological system of flora and fauna that exists in and around a river estuary.

Hard water Water that has a high mineral content. Hard water might contain calcium (Ca^{2+}) and magnesium (Mg^{2+}) ions, possibly other dissolved bicarbonates and sulphates and some metals.

Mantle The hot, interior region of the Earth's crust beneath the surface.

Mono pile A hollow steel cylinder that is sunk into the seabed to provide secure anchorage and a fixed operating platform.

Permo-Triassic sandstones Sandstone layers in the form of basins that are impervious to water flow.

Phytoplankton Animal life that exists in the sea at a microscopic level.

Radiothermal granites Granite rocks that contain heat as a result of the radioactive processes in the Earth's core.

Sedimentary basins Underlying rock basins impervious to water flow.

Spring tides For spring tides the height range between high and low tide is maximum. High tides are higher and low tides are lower than average water. **Slack water** time (the time between change in water direction) is shorter than average and tidal currents are stronger than average. **Neap tides** occur with less extreme variation in these conditions and the interval between spring and neap tides is around seven days.

Tectonically active Regions of the Earth's crust where the hot magma is close to the surface as a result of the geological processes (or tectonic processes) occurring in and below the crust.

Turbidity The cloudiness of or amount of suspended material in water.

USEFUL FUTURE SOURCES WEBSITES AND CONTACTS

Geothermal

British Geological Survey
www.bgs.ac.uk/
Kingsley Dunham Centre, Keyworth, Nottingham NG12 5GG
Tel: 0115 936 3100 Fax: 0115 936 3200

Geothermal Energy Association
www.geo-energy.org

Geological Survey of Ireland
www.gsi.ie
Beggars Bush, Haddington Road, Dublin 4, Republic of Ireland
Tel: +353 1 678 2000 Fax: +353 1 668 1782
LoCall: 1890 449900

Geothermal Resources Council
www.geothermal.org

continued overleaf

USEFUL FUTURE SOURCES WEBSITES AND CONTACTS *continued*

International Geothermal Association
http://iga.igg.cnr.it/index.php
Based in Italy, this association encourages research, development and utilization of geothermal resources worldwide.

US Department of Energy – Geothermal Technologies Program
www1.eere.energy.gov/geothermal/
Reports and descriptions of the technologies and projects.

Tidal Stream and Barrage/Wave Power

Atlas of UK Marine Renewable Energy Resources
www.berr.gov.uk/files/file27746.pdf
Provides access to maps of the tidal range and tidal stream resource around the UK coast.

British Wind Energy Association (Marine)
www.bwea.com/marine/index.html

The Carbon Trust – Marine Energy Challenge
www.thecarbontrust.co.uk/technology/technology-accelerator/marine_energy.htm
Technical guides on wave energy, tidal stream and wave power produced in 2004 and 2005.

EcoFys, Netherlands BV (The Wave Rotor)
www.ecofys.nl

Economic Viability of a Simple Tidal Stream Energy Capture Device (Department of Trade and Industry, 2007); available at www.berr.gov.uk/files/file37093.pdf

The Engineering Business Ltd (Stingray and the EB Frond)
www.engb.com

European Marine Energy Centre (EMEC) Ltd
www.emec.org.uk
Old Academy, Back Road, Stromness, Orkney KW16 3AW, Scotland
Tel: 01856 852060

Finavera Renewables (AquaBuOY)
www.finavera.com

Hammerfest Strøm AS
www.e-tidevannsenergi.com/index.htm
Norwegian company set up to develop tidal stream energy.

Marine Current Turbines (MCT) Ltd
www.marineturbines.com

Ocean Power Technologies (PowerBuoy)
www.oceanpowertechnologies.com

Rance Tidal Power Plant
www.edf.fr/html/en/decouvertes/voyage/usine/usine.html
Eléctricité de France website describing the world's most successful tidal barrage project.

Turning the Tide: Tidal Power in the UK (Sustainable Development Commission, 2007)
www.sd-commission.org.uk/pages/tidal.html
A detailed review of UK tidal power, cautiously supportive of a Severn Barrage project.

Pelamis Wave Power Ltd
www.pelamiswave.com

Sperboy (floating OWC device)
www.sperboy.com

Soil Machine Dynamics Ltd (SMD) (TidEL)
www.smd.co.uk

Wave Dragon
www.wavedragon.net

Wavegen (Limpet, Nearshore OWC)
www.wavegen.com

APPENDIX I

Getting Revenue from Renewable or Microgeneration Electricity

Large-scale developments, such as wind farms, major hydroelectric power stations and biomass community heating schemes, require to be brought forward on the basis of a sound financial investment with an appropriate business plan. The costs of these schemes will be highly site and plant dependent and returns will rely heavily on the market for the energy supplied. The economics and predicted returns for large-scale renewable energy projects with associated risks are beyond the scope of this book and should be evaluated in partnership with suitably experienced developers.

On the smaller scale, a simpler economic evaluation can be carried out and one of the most frequently used methods is simple payback. This is a calculation of how long it will take to recoup the cost of the initial investment from the savings made against energy bills. This will include, in some cases, revenue from the excess electricity generated and, additionally, revenue for the added environmental commodity value. In the UK this is obtained through the Renewable Obligation Certificates (ROCs) and Climate Change Levy Exemption Certificates (LECs) (see below). In larger-scale or community projects it may be possible to include revenue received for heat supplied.

The simple payback calculation is to divide the total cost of the equipment plus its installation by the amount saved annually on the energy bill. For example, and using illustrative figures only, a solar panel costing £2,500 that offset £100 per year from the energy bill (by providing heat that would otherwise have come from the immersion heater) would have a payback of twenty-five years. Obviously this will improve dramatically if a grant is available: a grant of £1,000, for example, will reduce the payback to fifteen years. These simple calculations do not take into account any increase in energy costs over the lifetime of the installation or a future cost levied against carbon (through a carbon tax on energy, for example). Payback also ignores the opportunity cost of using the money for another purpose, such as investing the money in a bank at 5 per cent interest. Clearly there are reasons, other than purely financial, for installing a renewable energy technology, for example to access the environmental benefit of environmentally friendly energy. Remember also that the lifespan of most renewable technology installations is greater than fifteen years. Payback is considered by some to be a very artificial method of assessing the true worth of renewable energy installations, particularly since investments such as double glazing and heating systems are rarely installed on this basis. Payback takes no account of the increase in the value of the property as a result of fitting the technology (and this is important in the new HIPS scheme) or of the potential to receive a renewable energy rates rebate or reduction in stamp duty which has been proposed.

In recent times with the cooling off of property prices, developers have been seeking to use the 'greenness' of a new house as a differentiator in an increasingly competitive property market. Clearly the route to lower-cost installations, concurrent with lower carbon buildings, is to ensure that the technologies are installed during construction. This will bring forward the economic benefits of economy of scale for bulk purchasing and site installation. It is possible that this may be encouraged through procurement legislation; this would be particularly effective for local authorities and councils who have responsibility for social housing.

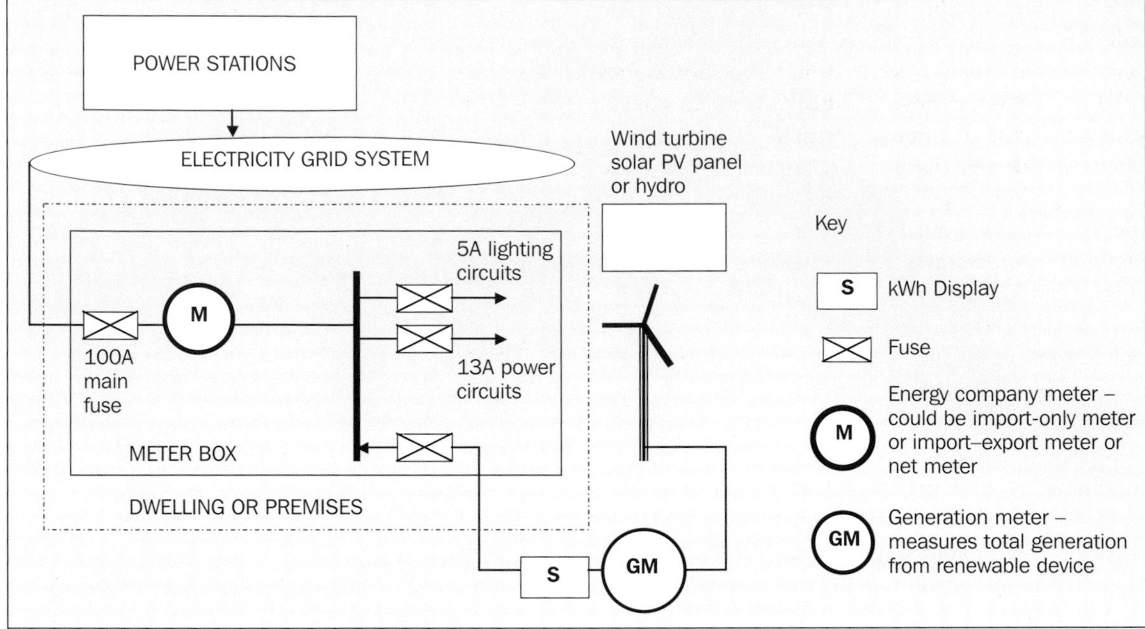

Figure 96: The renewable source of electricity generation is connected to the 'customer side' of the meter.

REVENUE FROM SPILLED ELECTRICITY

One of the most flexible and amenable sources of energy is electricity and, as we have seen, microgeneration devices, including solar photovoltaic (PV) cells, wind turbines and hydroelectricity plants all produce electricity. These devices can either be connected to the mains electricity system or they can be used to charge batteries in off-grid applications.

For buildings connected to the electricity system, electricity is conventionally provided (mostly from the fossil-fuelled power stations) by a connection to the grid through a cable, either overhead or under the ground, into a meter box. In the meter box, the grid supply passes through a main fuse to a consumer distribution board. The distribution board is where the various individually fused lighting and heating circuits in the building are connected to the mains. The electricity drawn from the grid is measured (in kilowatt hour units or kWh) and the cumulative amount used in the building is recorded on the energy supply company's main electricity meter. These meters are read, usually quarterly or monthly, by a meter reader from the energy supply company or sometimes by the customer. The readings show how much electricity has been consumed and so enable the electricity bill to be calculated. In the normal arrangement described above, electricity flow through the meter takes place in one direction only – that is, from the grid to the premises or dwelling – never in the other direction, unless there is a source of electricity generation on the customer side of the meter.

Wind turbines, solar powered photovoltaic (PV) panels or small-scale hydroelectric generators are directly connected to the electrical system of a building, but on the customer side of the main electricity meter (M in Figure 96). The electrical connection from these generators is made in the meter box. These devices can then provide electricity to run appliances and their electricity output might satisfy the total electrical demand when they generate. At such times electricity will not be taken from the electricity supply company, assuming the renewable generator output is greater than the electricity demand in the building. The intermittent nature of the output from the renewable technologies means that the electricity generated may vary on a minute-to-minute basis.

It is worth noting that the supply of electricity will only be available from the renewable technology when the wind blows (wind turbine), the Sun shines (PV) or the water flows (hydro). In circumstances where the cost of import electricity is higher than the renewable electricity export value, then the value can be maximized by consuming the renewable electricity when it is available. This means that washing machines, dishwashers and other electrical appliances should be used when the renewable generator is providing electricity. Unless the electricity generated can be used to charge batteries (or heat an electrical element in a hot water storage tank), then any electricity generated but not used will spill to the grid.

At other times the renewable technologies may not be generating sufficient output to provide the electricity requirement of the building. At these times the electricity supply company will provide the electricity as normal and electricity is delivered through the main electricity meter. In the case of most wind turbines and hydroelectric generators, the presence of the grid supply is necessary to enable an output from their generators (they are known as asynchronous generators), so the loss of the mains supply will mean that the renewable devices will not generate electricity.

Solar powered photovoltaic (PV) panels are different: they can provide an output when the grid is not present (rather like a battery) but the protection equipment they are fitted with (as part of the inverter) will shut them down to prevent generation occurring when the grid supply has been lost. There are situations, for example, when the electricity distribution network in an area has been isolated for repair or maintenance or when electrical work is taking place within the building. During these periods it would be undesirable to have renewable generation operating, since the electrical system could become live from the solar panel output, even though the grid had been purposefully or accidentally disconnected. This is a very dangerous situation and such renewable devices require protection to ensure they shut down rapidly and stay off-grid until the installation is completely safe and the mains supply is restored as normal.

Wind turbines, PV panels and hydro generators are provided with a meter, known as the generation meter (GM in Figure 96), that will record the electricity generated in kWh.

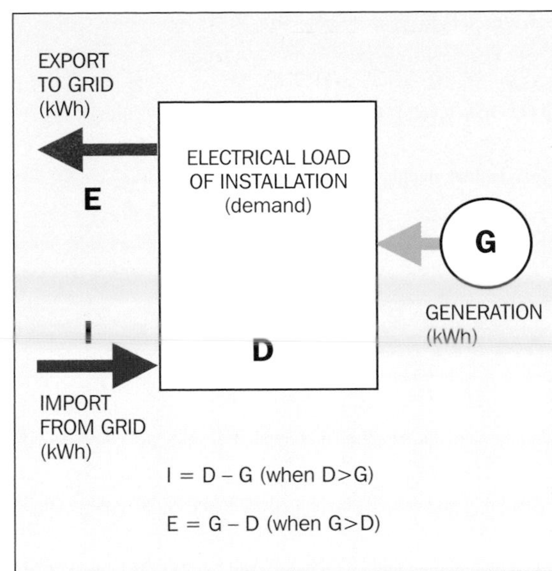

Figure 97: Import, export and generated electricity flows: imported electricity comes through the energy company's meter; exported electricity requires a new meter.

Referring to Figure 97, the following two scenarios outline the relationship between demand, generation, import and export:

1. When there is little or no renewable generation of electricity, but there is a high electrical demand from the building, this can be expressed as:

Imported electricity from the grid = electrical Demand – electricity from renewable Generator

$$I = D - G$$

2. A different situation exists when there is little or no electricity demand in the dwelling or premises, but there is a higher output from the renewable generator: for example at night when the wind is blowing, or during the day when there is bright sunshine supplying a PV panel but the occupants are at work or on holiday. At those times some of the renewable electricity generation is surplus and it will spill, that is it will be exported to the grid:

Export electricity onto the grid = electricity from Renewable Generator – electrical Demand

$$E = G - D$$

These scenarios constantly alternate as the building's electrical demand changes and the electricity output from the renewable generator fluctuates with the changing wind, water flow or solar illumination.

THE VALUE OF RENEWABLE ELECTRICITY

There are two attributes of the electricity from the renewable generation source that need consideration. The first is its value as a replacement for the conventional (fossil fuel-generated) electricity that would otherwise have been taken from the electricity supply company. The second is the environmental value of its contribution to the reduction in carbon, owing to its status as renewable electricity. In the UK this environmental value can be captured through Renewable Obligation Certificates (ROCs), Renewable Energy Guarantee of Origin Certificates (REGOs) and Levy Exemption Certificates (LECs).

From the electricity company's point of view, the electricity value will change depending on when it is generated and demanded. Electricity demanded at peak periods of the day, between 4.30pm and 7.00pm (particularly in winter) will have a high value, but electricity generated at other times (especially at night) may have a much lesser value. The precise value of the electricity at any given time is related to the price paid to the conventional power stations for the electricity they supply (from the trading pool) and to the cost of transmitting and distributing the unit of electricity to any given location. Some energy companies will pay a value related to or some fraction of this cost, and others may also acknowledge the additional embedded generation benefits from microgeneration (associated with electricity transmission and distribution losses). In 2008 Scottish and Southern Energy, for example, introduced an export tariff of 18p/kWh for electricity from PV systems.

Some renewable energy advocates suggest that the electricity produced from renewable technologies should be remunerated at a much higher rate, a feed-in tariff at around 50p per unit of electricity generated. This was the case in Germany, where high payments were made for electricity units delivered from PV panels, resulting in a rapid development of the national PV industry. There are proposals that microgeneration technologies in the UK could be paid for their output through a similar feed-in tariff arrangement. These proposals would significantly increase the value of electricity from renewable technologies and reduce payback periods.

ROCS, REGO AND LECS

The environmental value of the electricity may be more difficult to estimate, but most electricity companies will assign a value based on the market price of ROCs and LECs. Some energy companies, although not all, will offer agreements that effectively purchase both the surplus electricity and the environmental value of the output from the renewable generator (see below). A few companies will buy back the surplus at the same rate as they charge, a practice known as net metering. Conventional electricity meters used for billing cannot record spilled units put onto the grid, so a new type, known as an import-export meter, will be required. These meters have already been fitted by some energy companies, although they may attract an extra charge. Exporting electricity to the system is normally done using a half hour meter (HH meter). This means that the electricity exported is recorded in half hourly slots. These can be expensive. In Britain the rules are that generators up to 30kW will not require HH meters, but above that threshold they are necessary. The situation in respect of the metering of renewable generators is changing continually and it is recommended that a range of suppliers should be contacted to see who is offering the best deal.

Renewable Energy Certificates or ROCs are a central part of the UK's renewable energy support mechanism (the Renewable Obligation or RO). Each certificate (one ROC) is issued by the Office of Gas and Electricity Markets (Ofgem) annually in respect of every megawatt hour (MWh) of renewable energy generated. These certificates are auctioned in a 'ROC market': so far, ROCs have attracted a value of around £35 per MWh. Payment for ROC-accredited generation requires an approved generation meter (to register kWh generated); details of these are available from the local energy company.

In a revision of the UK RO, which takes effect from April 2009, renewable energy technologies will be banded to recognize the practical constraints experienced by the more mature technologies, such as onshore wind. The revision also helps to bring forward investment in the emerging technologies. Projects commissioned after July 2006 will receive the revised (generally higher) rates. Support for sewage gas and landfill gas generation, and co-firing using non-energy crop biomass, reduces to 0.25 ROCs per megawatt hour (MWh), although grandfathering (the maintenance of existing agreements in the Renewable Obligation) will protect those projects already approved by April 2009 and operational within two years thereafter. Established renewables, such as onshore wind and hydro, continue to receive single ROCs, as does co-firing using energy crops. Offshore wind and dedicated biomass power stations receive 1.5 ROCs, while emerging renewables such as anaerobic digestion, biomass gasification, energy crops for power generation, wave and tidal-stream power and photovoltaics get a substantial boost with double ROCs.

This move to boost the revenue from renewable energy projects should increase investor confidence, although there will still be doubts around the extent to which banks will lend money against the uncertain value of ROCs. Based on some predictions, double ROCs could be worth around £90 to £110/MWh on average, including the base price of electricity, but lenders may still consider them less preferable to a longer-term guaranteed feed-in tariff.

In the UK the Climate Change Levy (CCL) was introduced in April 2001 to encourage business energy users to become more energy efficient and to increase the deployment of renewable energy. Levy Exemption Certificates (LECs) can be used to offset any CCL payments required by a business. Both ROCs and LECs are paid on the total number of generated (the maintenance of existing agreements in the Renewable Obligation) units (not exported units) and they provide an additional source of income for electricity from renewable generation.

A further benefit to renewables is available in the form of REGOs. The Renewable Energy Guarantee of Origin certification scheme permits generators to

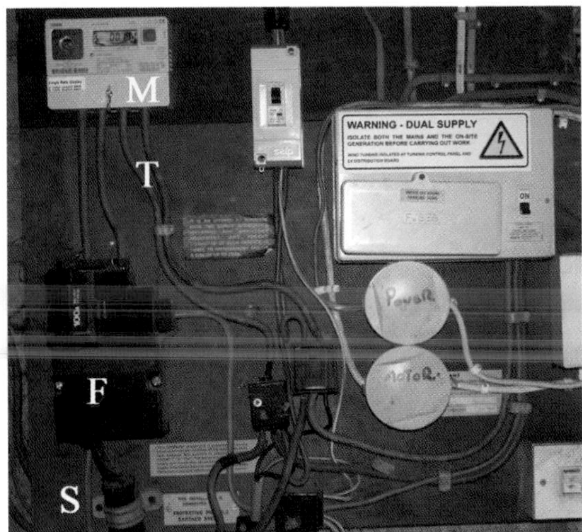

Figure 98: Typical meter board layout showing incoming grid supply (S), main fuse (F), import-export meter (M), tails to consumer unit (T) and other electrical connections.

identify and verify the source of their generation as environmentally sound. The scheme is available to all renewable energy generators, regardless of scale. This scheme provides the purchasers of renewable electricity with a guarantee that it is produced from recognized green sources.

Individuals can complete the Ofgem accreditation forms and get on the Ofgem registers as renewable generators. However, completing the necessary documentation to receive these certificates and secure the payments is recognized as excessively bureaucratic and complex. Once the certificates have been obtained, there are several options available that will enable their value to be realized. Some energy companies offer a sell and buy-back agreement, although certificate holders may wish to find a buyer or trade them in the market. Rather than get embroiled in the complexity of these operations, another option is to use a broker company, which will complete the forms and arrange for the certificates to be traded as necessary.

There are a growing number of energy companies selling electricity generated by renewable sources, such as large wind farms or hydroelectric stations. These are known as green offerings or green tariffs, and in some cases the electricity may attract a premium. (For a list of green suppliers, *see* pages 25–27.)

APPENDIX II

Renewable Energy Standards

There are a variety of standards applicable to renewable energy, including European, British Standards and IEC Standards. The examples listed below are particularly useful when designing a renewable energy system.

SOLAR THERMAL/SOLAR WATER HEATING

EN 12975-1:2006: Thermal solar systems and components – Solar collectors – Part 1: General Requirements
EN 12975-2:2006: Thermal solar systems and components – Solar collectors – Part 2: Test methods
EN 12976-1:2006: Thermal solar systems and components – Factory made systems – Part 1: General requirements
EN 12976-2:2006: Thermal solar systems and components – Factory made systems – Part 2: Test methods
EN ISO 9488:1999: Solar energy – Vocabulary (ISO 9488:1999)
ENV 12977-1:2001: Thermal solar systems and components – Custom built systems – Part 1: General requirements
ENV 12977-2:2001: Thermal solar systems and components – Custom built systems – Part 2: Test methods
ENV 12977-3:2001: Thermal solar systems and components – Custom built systems – Part 3: Performance characterization of stores for solar heating systems

HEAT PUMPS

CEN/TS 14825:2003: Air conditioners, liquid chilling packages and heat pumps with electrically driven compressors for space heating and cooling – Testing and rating at part load conditions
EN 12178:2003: Refrigerating systems and heat pumps – Liquid level indicating devices – Requirements, testing and marking
EN 12263:1998: Refrigerating systems and heat pumps – Safety switching devices for limiting the pressure – Requirements and tests
EN 12309-1:1999: Gas-fired absorption and adsorption air-conditioning and/or heat pump appliances with a net heat input not exceeding 70kW – Part 1: Safety
EN 12309-2:2000: Gas-fired absorption and adsorption air-conditioning and/or heat pump appliances with a net heat input not exceeding 70kW – Part 2: Rational use of energy
EN 13136:2001: Refrigerating systems and heat pumps – Pressure relief devices and their associated piping – Methods for calculation
EN 13136:2001/A1:2005: Refrigerating systems and heat pumps – Pressure relief devices and their associated piping – Method for calculation
EN 13313:2001: Refrigerating systems and heat pumps – Competence of personnel
EN 14276-1:2006: Pressure equipment for refrigerating systems and heat pumps – Part 1: Vessels – General requirements
EN 14276-2:2007: Pressure equipment for refrigerating systems and heat pumps – Part 2: Piping – General requirements
EN 15450:2007: Heating systems in buildings – Design of heat pump heating systems

EN 1736:2000: Refrigerating systems and heat pumps – Flexible pipe elements, vibration isolators and expansion joints – Requirements, design and installation

EN 1861:1998: Refrigerating systems and heat pumps – System flow diagrams and piping and instrument diagrams – Layout and symbols

EN 255-3:1997: Air conditioners, liquid chilling packages and heat pumps with electrically driven compressors – Heating mode – Part 3: Testing and requirements for marking for sanitary hot water units

EN 255-3:1997/AC:1997: Air conditioners, liquid chilling packages and heat pumps with electrically driven compressors – Heating mode – Part 3: Testing and requirements for marking for sanitary hot water units

EN 378-1:2008: Refrigerating systems and heat pumps – Safety and environmental requirements – Part 1: Basic requirements, definitions, classification and selection criteria

EN 378-2:2008: Refrigerating systems and heat pumps – Safety and environmental requirements – Part 2: Design, construction, testing, marking and documentation

EN 378-3:2008: Refrigerating systems and heat pumps – Safety and environmental requirements – Part 3: Installation site and personal protection

EN 378-4:2000/A1:2003: Refrigerating systems and heat pumps – Safety and environmental requirements – Part 4: Operation, maintenance, repair and recovery

EN 378-4:2008: Refrigerating systems and heat pumps – Safety and environmental requirements – Part 4: Operation, maintenance, repair and recovery

EN 814:1997: Air conditioners and heat pumps with electrically driven compressors – cooling mode (3 parts), BS1

ISO 13256-1:1998: Water-source heat pumps – Testing and rating for performance – Part 1: Water-to-air and brine-to-air heat pumps

ISO 13256- 2:1998: Water-source heat pumps – Testing and rating for performance – Part 2: Water-to-water and brine-to-water heat pumps

BIOFUELS

CEN/TS 14588:2004C: Solid biofuels – Terminology, definitions and descriptions

CEN/TS 14774-1:2004C: Solid biofuels – Methods for determination of moisture content – Oven dry method – Part 1: Total moisture – Reference method

CEN/TS 14774-2:2004C: Solid biofuels – Methods for the determination of moisture content – Oven dry method – Part 2: Total moisture – Simplified method

CEN/TS 14774-3:2004C: Solid biofuels – Methods for the determination of moisture content – Oven dry method – Part 3: Moisture in general analysis sample

CEN/TS 14775:2004C: Solid biofuels – Method for the determination of ash content

CEN/TS 14778-1:2005C: Solid biofuels – Sampling – Part 1: Methods for sampling

CEN/TS 14778-2:2005C: Solid biofuels – Sampling – Part 2: Methods for sampling particulate material transported in lorries

CEN/TS 14779:2005C: Solid biofuels – Sampling – Methods for preparing sampling plans and sampling certificates

CEN/TS 14780:2005C: Solid biofuels – Methods for sample preparation

CEN/TS 14918:2005C: Solid biofuels – Method for the determination of calorific value

CEN/TS 14961:2005C: Solid biofuels – Fuel specifications and classes

CEN/TS 15103:2005C: Solid biofuels – Methods for the determination of bulk density

CEN/TS 15104:2005C: Solid biofuels – Determination of total content of carbon, hydrogen and nitrogen – Instrumental methods

CEN/TS 15105:2005C: Solid biofuels – Methods for determination of the water soluble content of chloride, sodium and potassium

CEN/TS 15148:2005C: Solid biofuels – Method for the determination of the content of volatile matter

CEN/TS 15149-1:2006C: Solid biofuels – Methods for the determination of particle size distribution – Part 1: Oscillating screen method using sieve apertures of 3.15mm and above

Table 42: IEC wind turbine standards.			
Working Group	**Title**	**Purpose**	**Document Number**
WG-1 WG-2 WG-3	Safety Requirements for Large Wind Turbines	Principal standard defining design requirements	IEC 1400-1*
WG-4	Small Wind Turbine Systems	Principal standard defining design requirements for small turbines	IEC 1400-2*
WG-5	Acoustic Emission Measurement Techniques	Defines acoustic measurements methods	IEC 1400-11*
WG-6	Performance Measurement Techniques	Defines performance measurement techniques	IEC 1400-12*
WG-7	Revision of IEC 1400-1	Edition 2 of IEC 1400-1	1400-1 Ed2
WG-8	Blade Structural Testing	Defines methods for blade structural testing	1400-23
WG-9	Wind Turbine Certification Requirements	Defines certification requirements (harmonized version of several European standards)	1400-22
WG-10	Power Quality Measurements	Defines power quality measurement techniques	1400-21
WG-11	Structural Loads Measurement	Defines methods for measuring operational loads	1400-13
*Published Standard			

CEN/TS 15149-2:2006C: Solid biofuels – Methods for the determination of particle size distribution – Part 2: Vibrating screen method using sieve apertures of 3.15mm and below
CEN/TS 15149-3:2006C: Solid biofuels – Methods for the determination of particle size distribution – Part 3: Rotary screen method
CEN/TS 15150:2005C: Solid biofuels – Methods for the determination of particle density
CEN/TS 15210-1:2005C: Solid biofuels – Methods for the determination of mechanical durability of pellets and briquettes – Part 1: Pellets
CEN/TS 15210-2:2005C: Solid biofuels – Methods for the determination of mechanical durability of pellets and briquettes – Part 2: Briquettes
CEN/TS 15234:2006C: Solid biofuels – Fuel quality assurance
CEN/TS 15289:2006C: Solid biofuels – Determination of total content of sulphur and chlorine
CEN/TS 15290:2006C: Solid biofuels – Determination of major elements
CEN/TS 15296:2006C: Solid biofuels – Calculation of analyses to different bases
CEN/TS 15297:2006C: Solid biofuels – Determination of minor elements
BS EN 303-5:1999: In addition to specifications for biomass fuels, there are also specifications for biomass combustion equipment. BS EN 303-5:1999 applies to heating boilers for solid fuels, hand and automatically fired, nominal heat output of up to 300kW. It is the local UK implementation by the BSI of EN 303-5

WIND TURBINES

EN 45510-5-3:1998: Guide for procurement of power station equipment – Part 5-3: Wind turbines
BS EN 61400-12-1: 2006: Wind turbines: Power performance measurement of grid connected wind turbines
BS EN 61400-11: 2003: Wind turbine generator systems: Acoustic noise measurement techniques
BS EN 61400-2: 2006: Wind turbine generator systems: Design requirements of small wind systems
ICS: 27.180: Wind turbine systems and other alternative sources of energy

INTERNATIONAL ELECTROTECHNICAL COMMISSION (IEC) STANDARDS

The IEC is the world's leading organization for the preparation and publication of International Standards for all electrical, electronic and related technologies, collectively known as 'electrotechnology'. IEC Standards cover a vast range of technologies from power generation, transmission and distribution to home appliances and office equipment, semiconductors, fibre optics, batteries, nanotechnologies, solar energy and marine energy converters.

Wind Energy

A number of IEC Standards have been created or are in draft. (For further details, see http://www.awea.org/standards/iec_stds.html#WG123.)

Solar PV

IEC 60891 (1987-04): Procedures for temperature and irradiance corrections to measured I-V characteristics of crystalline silicon photovoltaic devices
IEC 60891-am1 (1992-06): Amendment 1 – Procedures for temperature and irradiance corrections to measured I-V characteristics of crystalline silicon photovoltaic devices
IEC 60904-1 (2006-09): Photovoltaic devices – Part 1: Measurement of photovoltaic current-voltage characteristics
IEC 60904-1 (2006-09): Photovoltaic devices – Part 1: Measurement of photovoltaic current-voltage characteristics
IEC 60904-2 (2007-03): Photovoltaic devices – Part 2: Requirements for reference solar devices
IEC 60904-3 (1989-02): Photovoltaic devices – Part 3: Measurement principles for terrestrial photovoltaic (PV) solar devices with reference spectral irradiance data
IEC 60904-5 (1993-10): Photovoltaic devices – Part 5: Determination of the equivalent cell temperature (ECT) of photovoltaic (PV) devices by the open-circuit voltage method
IEC 60904-7 (1998-03): Photovoltaic devices – Part 7: Computation of spectral mismatch error introduced in the testing of a photovoltaic device
IEC 60904-8 (1998-02): Photovoltaic devices – Part 8: Measurement of spectral response of a photovoltaic (PV) device
IEC 60904-9 (2007-10): Photovoltaic devices – Part 9: Solar simulator performance requirements
IEC 60904-10 (1998-02): Photovoltaic devices – Part 10: Methods of linearity measurement
IEC 61173 (1992-09): Overvoltage protection for photovoltaic (PV) power generating systems – Guide
IEC 61194 (1992-12): Characteristic parameters of stand-alone photovoltaic (PV) systems
IEC 61215 (2005-04): Crystalline silicon terrestrial photovoltaic (PV) modules – Design qualification and type approval
IEC 61277 (1995-03): Terrestrial photovoltaic (PV) power generating systems – General and guide
IEC 61345 (1998-02): UV test for photovoltaic (PV) modules
IEC 61646 (1996-11): Thin-film terrestrial photovoltaic (PV) modules – Design qualification and type approval
IEC 61683 (1999-11): Photovoltaic systems – Power conditioners – Procedure for measuring efficiency
IEC 61701 (1995-03): Salt mist corrosion testing of photovoltaic (PV) modules
IEC 61702 (1995-03): Rating of direct coupled photovoltaic (PV) pumping systems
IEC 61724 (1998-11): Photovoltaic system performance monitoring – Guidelines for measurement, data exchange and analysis

IEC 61725 (1997-05): Analytical expression for daily solar profiles

IEC 61727 (2004-12): Photovoltaic (PV) systems – Characteristics of the utility interface

IEC 61730-1 (2004-10): Photovoltaic (PV) module safety qualification – Part 1: Requirements for construction

IEC 61730-2 (2004-10): Photovoltaic (PV) module safety qualification – Part 2: Requirements for testing

IEC 61829 (1995-03): Crystalline silicon photovoltaic (PV) array – On-site measurement of I-V characteristics

IEC/TS 61836 (2007-12): Solar photovoltaic energy systems – Terms, definitions and symbols

IEC 62093 (2005-03): Balance-of-system components for photovoltaic systems – Design qualification natural environments

IEC 62108 (2007-12): Concentrator photovoltaic (CPV) modules and assemblies – Design qualification and type approval

IEC 62124 (2004-10): Photovoltaic (PV) stand-alone systems – Design verification

IEC/TS 62257-1 (2003-08): Recommendations for small renewable energy and hybrid systems for rural electrification – Part 1: General introduction to rural electrification

IEC/TS 62257-2 (2004-05): Recommendations for small renewable energy and hybrid systems for rural electrification – Part 2: From requirements to a range of electrification systems

IEC/TS 62257-3 (2004-11): Recommendations for small renewable energy and hybrid systems for rural electrification – Part 3: Project development and management

IEC/TS 62257-4 (2005-07): Recommendations for small renewable energy and hybrid systems for rural electrification – Part 4: System selection and design

IEC/TS 62257-5 (2005-07): Recommendations for small renewable energy and hybrid systems for rural electrification – Part 5: Protection against electrical hazards

IEC/TS 62257-6 (2005-06): Recommendations for small renewable energy and hybrid systems for rural electrification – Part 6: Acceptance, operation, maintenance and replacement

IEC/TS 62257-7-1 (2006-12): Recommendations for small renewable energy and hybrid systems for rural electrification – Part 7-1: Generators – Photovoltaic arrays

IEC/TS 62257-8-1 (2007-06): Recommendations for small renewable energy and hybrid systems for rural electrification – Part 8-1: Selection of batteries and battery management systems for stand-alone electrification systems – Specific case of automotive flooded lead-acid batteries available in developing countries

IEC/TS 62257-9-2 (2006-10): Recommendations for small renewable energy and hybrid systems for rural electrification – Part 9-2: Microgrids

IEC/TS 62257-9-3 (2006-10): Recommendations for small renewable energy and hybrid systems for rural electrification – Part 9-3: Integrated system – User interface

IEC/TS 62257-9-4 (2006-10): Recommendations for small renewable energy and hybrid systems for rural electrification – Part 9-4: Integrated system – User installation

IEC/TS 62257-9-5 (2007-06): Recommendations for small renewable energy and hybrid systems for rural electrification – Part 9-5: Integrated system – Selection of portable PV lanterns for rural electrification projects

IEC/TS 62257-12-1 (2007-06): Recommendations for small renewable energy and hybrid systems for rural electrification – Part 12-1: Selection of self-ballasted lamps (CFL) for rural electrification systems and recommendations for household lighting equipment

IEC/PAS 62111 (1999-07): Specifications for the use of renewable energies in rural decentralized electrification

Index